I USED TO KNOW THAT
GEOGRAPHY

STUFF YOU FORGOT FROM SCHOOL

WILL WILLIAMS

FOREWORD BY CAROLINE TAGGART

Michael O'Mara Books Limited

This paperback edition first published in 2017

First published in Great Britain in 2010 by
Michael O'Mara Books Limited
9 Lion Yard
Tremadoc Road
London SW4 7NQ

A CIP catalogue record for this book is available from the British Library.

Papers used by Michael O'Mara Books Limited are natural, recyclable
products made from wood grown in sustainable forests. The
manufacturing processes conform to the environmental regulations of the
country of origin.

ISBN: 978-1-78243-755-0 in paperback print format
ISBN: 978-1-84317-939-9 in Mobipocket format
ISBN: 978-1-84317-940-5 in ePub format

1 2 3 4 5 6 7 8 9 10

Illustrations by David Woodroffe
Designed and typeset by glensaville.com

Printed and bound by CPI Group (UK) Ltd, Croydon, CR0 4YY

www.mombooks.com

I USED TO KNOW THAT
GEOGRAPHY

CONTENTS

FOREWORD

When the original version of *I Used to Know That* was published, I spent a very jolly couple of days in a small BBC studio in central London. With headphones over my ears and a microphone in front of me, I talked to people on radio stations all over the country about the book: why I had written it, what they liked about it and what brought back hideous memories.

To my surprise, the hideous memories were what excited people most. Top of the list – and this bit *wasn't* a surprise – was maths. One listener said that just looking at the letters $a + b = c$ on the page had brought him out in a cold sweat, even though he no longer had any idea why. Another radio station carried out a series of interviews in the street asking people, among other things, if they knew who Pythagoras was. 'Oh yes,' said one man, 'he's to do with triangles and angles and all that malarkey.'

I thought that was wonderful: 'all that malarkey' summed up perfectly the way many of my generation were taught. We had to learn it (whatever 'it' was); we were never really told why; and, once exams were over, unless we went on to be engineers or historians or something, we never thought about it again. But it lingered somewhere at the back of our minds, which may be why *I Used to Know That* touched a chord.

However, covering five major subjects and including a catch-all chapter called General Studies meant that a single small volume couldn't hope to deal with anything in much depth. This is where the individual titles in this series come in: if *I Used to Know That* reminded us of things that we learned once, these books will expand on them, explain why they were important and even, in the case of geography, update us on theories that have been dismissed, developed or expanded upon since we went to school. If you enjoy this one, look out for *I Used to Know That: English, Maths, History* and *General Science* as well.

The teaching of geography has changed beyond recognition over the last few decades. I was of what Will Williams calls the 'capes and bays' generation: we learned the names of places and the heights of mountains, but it never crossed anyone's mind to take us out of doors to stroll along a beach or wade through a river, to see for ourselves how these things actually worked. And certainly no one ever persuaded me that geography was fascinating because it was all around me, an unavoidable and ever-changing part of my daily life – and the daily life of everyone else in the world.

Will Williams brings the subject alive principally by showing just how wide-ranging it is. Everything from volcanic eruptions to ecotourism, climate change to models for the development of settlements is here – and it is all geography! As Will says, this is a holistic discipline, encompassing science, economics and sociology, not to mention the geographical sub-disciplines of geology, geomorphology, tectonics and others too numerous to mention. Even if you are young enough to have been taught the theory of how landscape has changed over time, or the economic and social importance of population growth, you are sure to find new insights in this book; if you never got further than memorizing the lengths of the Nile, the Amazon and the Congo, it will be a revelation.

In other words, whether it is a trip down memory lane, a voyage of discovery or a helpmeet for future quizzes, *I Used to Know That: Geography* has something to offer anyone with an interest in the workings of this planet and the people who live on it. And if that sounds like a sweeping claim – well, that's geography for you.

CAROLINE TAGGART

INTRODUCTION

When you took off for the annual 'Family Summer Holiday', it would be your father who would navigate: navigate *and* drive. You and your sibling(s) would fight in the back of the car and your mother took sole responsibility for 'The Map'. This meant that when you needed a detour or (whisper it) got lost, it was Mum who, unfairly, would be to blame. Not that Dad ever looked at the map; he preferred the method of learning the roads and the sequence of settlements en route. Nowadays, folks just plug in the destination location, set the satnav and off they go, hopefully avoiding low bridges and dead ends.

This vignette encapsulates the role of geography in everyday life and unfortunately demonstrates the limits of its reach into many people's lives. Avoiding all of the talk about how the world of work has changed and how we have all become more isolated from people in our own communities: just think of the maps! Be it Lewis and Clark in the USA, Flinders across Australia or Livingstone in Africa, the great explorers didn't set out to create maps for us to then downgrade them in the face of technology.

Maps are where most people first encounter geography and though satnav demonstrates the limits of people's engagement now, maps have made a pretty spectacular comeback. Modern geographers go nuts over 'geographical information systems' (GIS) and you too may, probably unwittingly, have become a geographer at least once in your working day. The Internet is awash with maps: maps with data on them, maps that show you where your friends (or at least their mobile phones) are, maps that show you when your house will flood, maps that locate your nearest restaurant, maps in fact that can show anything and everything. So geography is here to stay, a vital part of all our lives.

To be a geographer in the opening decades of the twenty-first century is to be on the one hand excited about the endless possibilities for travel, study and fulfilment, yet on the other to be frustrated with the lack of true joined-up thinking out there. Geography has a unique and valuable role to play in bringing together the strands that surround complex issues and produce clarity of focus. Nowhere can this be seen in more sharp relief than in the debate over climate change.

Across the world we know that use of renewable resources must be a foundation for our descendants. We know too that, locally, weather patterns have changed over time as the climate has varied in the past. Also, we know that carbon dioxide levels have rocketed upwards since we have helped move carbon from its stores in the ground, up into the atmosphere. But we don't *actually* know that the climate is changing because of man. It probably is, but it doesn't matter. The reality is that due to dwindling supplies of fossil fuels sometime in the future, we will have to change our reliance.

And it's geography that plays a part in all facets of this debate, and geographers who are perhaps uniquely placed to spot the simple coherent pathway to explanation. Be it economic concerns over the rising price of oil, environmental concerns over the impact of fossil fuel production and combustion, scarcity concerns where national supplies will be cut off or political concerns over one country's influence on others – it doesn't matter. In the end we need to become more sustainable, hence we need to adapt to renewable resources and we need geographers to bring together the disparate fields of enquiry to provide the ideas for moving to the next stage of development.

This issue is our modern 'Malthusian debate' (*see* p.117), that cornerstone of public consciousness that yields column inches of erudite copy and its fair share of mumbo-jumbo too. We now have the twenty-four-hour news network and the live blogosphere to keep the debate swirling around the world.

Who would have thought that when you were learning your US state capitals, your longest rivers in the world and your flags of the UN you were laying down the foundations for a subject that would become more and more relevant as the world has grown in complexity?

THE PHYSICAL

WORLD

☾

RIVERS

As rivers provide invaluable resources for so many people around the world – and take the lives of many too – this is probably a good place to start investigating just what we remember of our physical geography. A river – water flowing in a channel downhill – is simple to understand: the merry cascade tumbling down a mountainside; the awe-inspiring waterfall; the long, slow, meandering waterway; the huge body of water of a river in flood, bearing down on all before it . . .

You may have learned where they were and how long they were, or you may have waded into them and measured them, pebbles and all. Disappointingly, the former tended to be those in exotic places like Egypt, while the latter were usually whatever waterway could be found locally.

THE LONG PROFILE

This refers to the 'make-up' of the river, how it changes shape from source to mouth. Rivers are usually divided into three sections: the upper course, the middle course and the lower course, and they can start from springs, bogs or run-off from the sides of steep mountains, which often get rainfall to sustain the streams. Moving from source to mouth, the character of most rivers changes significantly as a result of the interaction of three factors:

* the rate at which the altitude of the riverbed decreases with distance (the gradient).
* the relationship between the friction surface of the bed and banks and the cross-sectional area of the channel (channel morphology).
* the small-scale features of the riverbed and their impact on the way the water flows, eddies and tumbles (bed roughness).

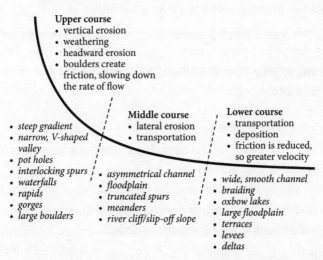

Upper course
• vertical erosion
• weathering
• headward erosion
• boulders create friction, slowing down the rate of flow

Middle course
• lateral erosion
• transportation

Lower course
• transportation
• deposition
• friction is reduced, so greater velocity

• *steep gradient*
• *narrow, V-shaped valley*
• *pot holes*
• *interlocking spurs*
• *waterfalls*
• *rapids*
• *gorges*
• *large boulders*

• *asymmetrical channel*
• *floodplain*
• *truncated spurs*
• *meanders*
• *river cliff/slip-off slope*

• *wide, smooth channel*
• *braiding*
• *oxbow lakes*
• *large floodplain*
• *terraces*
• *levees*
• *deltas*

Obviously, the work of water on the land is only half the story: the underlying rock provides the sketch pad on which the water draws.

WHERE THE RIVER RUNS FASTEST

You might think it's logical for the river to flow fastest at the source in the upper course, with that steep gradient? Well, you'd think so and it is certainly true that most waterfalls are in the upper course and, yes, they are running rather quickly. But the answer is more complex than that. Over a significant section of river the fastest velocities will be found where the influence of the gradient is enough to defeat the dark forces of friction. In the end, it is in the lower course, where the river channel is most efficient, that the average velocity is at its highest.

HOW DO YOU MEASURE THE WATER IN A RIVER?

The volume of water flowing in a river is called its discharge, and that is calculated as cross-sectional area multiplied by average velocity.

From this derive the units for measuring discharge: cumecs – **cu**bic **m**etres per **se**cond. The universally utilized symbol for river discharge is Q.

Shaping a Landscape

Rivers give us a good illustration of the importance of 'high-magnitude-low-frequency events' in shaping a landscape. For most of the year, a river is easily able to transport the water in it without having too much energy left for erosion. Perhaps four or five times a year the discharge of the river will be enough to fill the channel (known as 'bankfull discharge'). At about this point the river will have the most energy it can handle given the current shape of the channel. But if the level of water exceeds this, the river will flood and instantly start to slow down. Hence the river can control the times of flood by changing the shape and long profile – by increasing or decreasing the erosion of the channel. But it can only do this if it can no longer transport the river discharge in the current channel. Therefore, only when the river is at the highest energy state it can handle will it start to do some new work.

How Rivers Do Their Work

We all learned the same mnemonic for the processes of fluvial erosion (and, it has to be said, coastal erosion too) – CASH:

Corrasion – rocks rubbing against the bed and banks

to alter the channel shape. Should of course really be called 'abrasion' but that would have left a problem for the mnemonic makers.

Attrition – rocks in the stream rubbing against each other to produce more rounded and smaller particles.

Solution – particles in the water dissolving into the river. In limestone areas with slightly acidic water this helps to produce some of the most luxuriously curvaceous features on our planet. Rainwater reacts with carbon dioxide as it falls through the air, making it weak carbonic acid. Even without any other atmospheric pollutants, rain will always be more acidic.

Hydraulic action – the force of the water in the channel against the bed or banks can cause air to become trapped, with the pressure weakening the bank and causing it to wear away.

How Rivers Move Load

The 'load' in question is, of course, all the material – gravel, pebbles, rocks – that's carried by the natural flow of the river. Because the river moves it has kinetic energy, and it uses that energy to do the following things:

Flowing – literally moving the water and no more. Apart from some solution on the way, the river will be doing little else.

Transporting – if there is enough energy then the water will carry material with it and move it downstream.

Eroding – apart from the particles involved in solution, there will be very little river erosion done unless the river is in a high-energy state. This may only happen in particular places of fast flow like rapids, or at certain times of the year when discharge is high.

TRANSPORT METHODS

Depending on the size of the particles and the velocity of the water, the river will use the following processes to transport material:

Solution – as you'd expect, soluble material dissolves in the river and moves with the water – an easy starter.

Suspension – fine particles such as mud or silt are suspended in the flow of the water and carried along.

Saltation – due to an increase in river energy, material is moved along in a sequence of temporary suspensions and depositions.

Traction – when a boulder is too big to be carried, or the flow of the water is too slow to entrain it, it is transported by traction – the simple rolling or sliding of load over the riverbed.

In reality, material will move down a river using a

variety of these processes. Traction load, for example, may find itself gathering speed, colliding with bed load and, for a brief moment, be lifted up into the suspended load. Chances are, of course, that for most load in most rivers, nothing much will happen for most of the year.

Hjulström's Curve

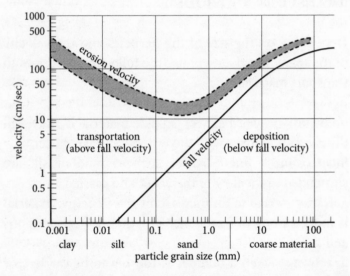

For geographers, Hjulström's Curve is one of the most fascinating graphs. In essence, it shows that the relationship between particle size and water velocity isn't quite as simple as it might seem. Created by Filip Hjulström, a Swedish geographer of the mid-twentieth century, it was produced from research in an artificial river called a flume. The flume allowed Hjulström to make small changes in the speed of the water flow and to record the impact on bed load of different sizes. The graph plots two lines. The higher line shows the speed of water flow needed to pick up a particle of a given size, and the lower line shows the minimum speed required to keep the particle in suspension in the water. Hence a particle will be deposited if the river slows down below the deposition speed and it will be picked up again if the river flows as fast as the entrainment or erosion speed. One can understand why a large rock may need a high velocity of water to pick it up (entrain it). One can also understand that if the water around the rock slows down or is replaced by turbulent slower water, the rock will drop. But why does it take so much energy to entrain the finest clay particles? After all, a piece of rock smaller than a grain of sand doesn't weigh that much.

The answer is both complex and simple. The simple bit is that fine particles pack closely together on the riverbed and water may not be able to penetrate this tight packing to lift the load. The complex part is that clay has a layered structure – a micelle. The clay particle is negatively charged, and this charge may help the particle to adhere to the bed and again increase the velocity of water flow needed to entrain it. Once entrained, the ultra-fine load will theoretically never be deposited, as its mass is so low that keeping it in suspension should require negligible energy.

It is load of this size that gives glacial rivers their cloudy appearance and makes the Amazon 'stripy' where it is formed at the confluence of the blackwater Rio Negro and the Rio Solimöes. Of course, the fine material cannot remain suspended for ever and has eventually to be deposited. It is this material that helps create landforms at the river's mouth.

MANNING'S EQUATION

In the latter half of the nineteenth century, the Irish engineer Robert Manning, seeking to further develop an existing mathematical method for estimating river velocity, produced this eponymous formula. With civil engineering projects getting bigger and bolder it was important that a method could be used to calculate river velocity. Manning's equation unites the three influences on river velocity in one equation:

Velocity = [Cross-sectional area/Wetted perimeter]$^{2/3}$ × Gradient$^{1/2}$ / Manning's n

Or

$$Velocity = \frac{R^{2/3} \cdot S^{1/2}}{n}$$

This theoretical model brings together the variables that influence the velocity of river flow. The cross-section and wetted perimeter give an indication of the efficiency of the channel shape, the gradient relates to the force acting down the long profile and 'Manning's n' is there to show the influence of the size of the load (bed roughness). The value can be worked out by sampling the sizes of the bed load, and, broadly speaking, the larger the bed load, the higher the value. So in the cold headwaters of Snowdonia a value of 0.10 can be estimated, whereas 0.02 will find you near the river's mouth. This goes some way to explaining why the fastest river flow does not occur in the upper course. (*See* 'Where the River Runs Fastest', p.18.) Near the source, gradient is high but the bed load is very large and the creation of riffles and pools reduces average velocities.

RIVER COMPETENCE AND CAPACITY

With variations in the river's ability to do work there are two descriptive phrases that are used:

River competence – measured by the largest individual load that can be entrained by the river at a given time.

River capacity – the total volume of load that can be carried by a river at any one time.

RIVER LANDFORMS

From waterfalls to deltas, oxbow lakes to meanders, river landforms are the features we love to remember.

WATERFALL FORMATION

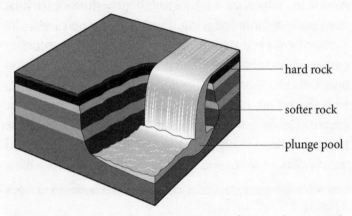

hard rock

softer rock

plunge pool

The kinetic energy in the water may first make it ripple over a change in gradient – and that's when the fun starts. When you consider that 1 cubic metre of water, carrying no load, weighs 1 tonne, you can see how it is going to do some serious damage to the riverbed and banks. Add to that the weaknesses within and between rock types, and the water begins to carve. The CASH processes (*see* p.19) at the base of the fall will lead to undercutting, and the removal of vertical support by the creation of the plunge pool will lead to occasional catastrophic collapses and the retreat and often deepening of the feature.

If you have a neat sandwich of weak rocks (especially of the sedimentary type) and harder rocks then the resulting gorge can be spectacular. The rates of gorge cutting by this process vary greatly, from 1 centimetre per year in the UK to nearly 1 metre per year for North American falls such as Niagara, which shows just how dynamic the forces of erosion and weathering can be.

Two other ways for waterfalls to form are by glaciers cutting deep valleys – leaving hanging valleys – and by uplift. In this latter case, if the land is moved upwards by tectonics (an earthquake, for example) faster than the rate of erosion lowers the rocks, then there will be a step in the landscape and any river will have to 'fall' over this step. The best of these waterfalls are fed by glaciers and can be seen on the south coast of Iceland – for example Seljalandsfoss (100 m drop).

THE WATERFALL TOP TEN

Let us not get too het up about this list. With many waterfalls you either end up measuring a series of steps or one long vertical drop. Through researching the following I became sadly disillusioned with the level of pedantry surrounding this subject, so I've gone for a consensus list.

Name of falls	Overall length of falls
Angel Falls, Venezuela	979 m
Tugela Falls, South Africa	948 m
Las Tres Hermanas, Peru	914 m
Olo'upena Falls, USA	900 m
Yumbilla Falls, Peru	896 m
Vinnufossen, Norway	860 m
Balåifossen, Norway	850 m
Pu'uka'oku Falls, USA	840 m
James Bruce Falls, Canada	840 m
Browne Falls, New Zealand	836 m

Rapids

Loved by canoeists and rafters alike, these are in fact just waterfalls with no obvious falls. Instead, debris-filled river channels with a small break in gradient lead to a large, frothy torrent of water. One way to experience this feature in full is to go white-water rafting on the Tully River in Queensland, Australia.

Riffles and Pools

In the upper course of the river, the channel is narrow and, normally, the bed load that has rolled down from the valley sides is large. The result is that there is often a sequence of debris dams that pond the water back into still pools, followed by water cascading over the debris –

the riffles. Taken over a large section, the average water velocity of these two landforms combined is very low.

THE FLOODPLAIN

Water in a river flows largely at below the capacity of the channel, but as we saw earlier, there are times when the river discharge fills the channel 'bankfull'. If the river is being asked to transport too much water then it will exceed bankfull and flood. Floodplains formed naturally by seasonal variations in the discharge of the river will have layers of fine river silt deposited on them. As the river spills over the banks the water is slowed down by friction. The result is that coarse material will be dropped first (beginning levee formation). This will leave flood water trapped on the floodplain and unable to drain back into the river. This natural phenomenon may be exaggerated by man-made levees that are very high and lead to more prolonged flooding than is natural. At the edge of the floodplain will be the 'bluff line', a band of higher ground that does not get contemporary flooding.

BRAIDING

You cannot watch any of Peter Jackson's *Lord of the Rings* trilogy without seeing this landform at its most spectacular. When a river full of load suddenly loses gradient and velocity, it deposits its load rapidly. This creation of a rock-choked valley often means the river has to split up to get down the unusually flat section. This splitting up gives rise to multiple channels and is termed 'braiding'. It is particularly stunning where historic glacial action has deposited a lot of material that the more recent river cannot hope to move, as may be found in parts of New Zealand where filming took place, notably in alpine environments such as that country's Southern Alps.

THE MEANDER

The development of a meander

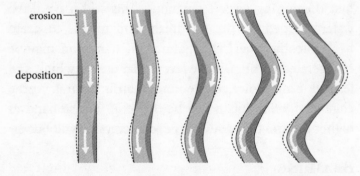

A classic of river landforms, found in the middle and lower course, the meander needs two basic ingredients: accelerating and decelerating water, coupled with helicoidal flow (the corkscrew-like motion of water as it flows along a river channel). When water is forced to

Cross-section of a meander

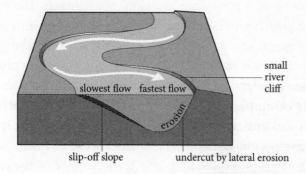

small
river
cliff

slowest flow   fastest flow

erosion

slip-off slope   undercut by lateral erosion

speed up over a narrower or shallower channel, it does. The water does not flow uniformly across the channel but is fastest on the outside of any helicoidal flow. The core of fastest-flowing water in the channel is deflected by the acceleration to one side of the bank or the other. Where this fast-flowing water gets pushed against the outside bank it will start to erode it, making the outer channel deeper and more efficient – so the water travels with more kinetic energy. Meanwhile, on the inside of the meander the water is now moving in a very inefficient channel, meaning that energy is lost, velocity drops and load is deposited – one bank is worn away, the other is filled in. The whole meander may move downstream over time, depending on the energy of the river and the relative strength of the bedrock.

THE OXBOW LAKE

Take one highly sinuous river, allow enough time for infrequent and high-energy events to occur and you eventually get an oxbow lake. As the river snakes along, a meander might curve so much that a U-shape is formed and the top parts of the 'U' work towards each other – in a process engagingly known as 'necking' – until the river cuts across the narrow neck of land between them. Now with a straighter path and usually more velocity, the river deepens its new short channel so that the entrance and exit of the meander loop are left high and dry, blocked by deposited load, leaving a crescent-shaped – oxbow – lake. After time there may not even be a lake left and the relic shapes in the landscape remain the only clues to the river's past.

AT THE MOUTH

When two rivers meet we pass it by as a 'confluence', but when it meets the sea it forms a number of things. We arrive at the French word *fleuve*, a river emptying into an ocean. The shape of the coastal/river interface owes much to the geology and the energy of the environment.

The Delta

A low-lying landform consisting mainly of silt and mud laid down by a river at its mouth, the delta owes its name to its resemblance to the Greek letter (Δ). There are three distinctly different deltas:

The arcuate delta – seen at the Nile in Egypt, the triangular shape of this type of delta makes it the most recognizable.
The cuspate delta – usually has few distributaries (think tributaries but the other way round) branching off. More like a coastal river floodplain. Can be seen at the mouth of the Tigris.
The New Orleans or bird's foot delta – a branching morphology where levees on each distributary allow them to extend out beyond any arcuate delta shape.

While each of these landforms differs, they all share some basic ingredients. Although fine sediments carried by a river should *theoretically* never be deposited (*see* Hjulström's Curve, p.122), at the coast they are. Fine particles have an electrostatic charge on them that makes them repel each other in fresh water and hence stay small. The salts in sea water serve to attract the particles and they thus form a larger aggregate. As the river has now run into the sea (or lake), it has slowed down dramatically and waterborne particles settle to the bottom.

Over a year the discharge of the river will vary – glacier-fed Rhône, rain-fed Severn, snowmelt Colorado – they all experience periods of low and higher flow. This leads to differences in the load carried throughout the year and results in layers or strata in the delta – each one an annual record of the work the river has done. As long as tidal and coastal processes do not remove more sediment than is deposited by the river, then the delta will grow.

ESTUARIES

By comparison with the spectacular delta landscapes, estuaries tend to be overlooked. This is strange as the majority of rivers meet the sea at estuaries. The estuary represents a zone where at low tide the river is able to flow out to sea with a basic small channel. At high tide, though, the river can no longer flow out to sea as the sea itself is blocking the route. Hence the fresh water of the river meets and mingles with the salty water of the sea and deposition occurs. This mud is full of nutrition and food for wading birds in particular. It is also good for shellfish – but not so good for anyone unwittingly surrounded by a rising tide.

RIAS

Also known as a drowned river valley, a ria (the Spanish word for estuary) is formed where sea levels rise and the valley previously at sea level is submerged. Change over time is one of the great themes in geography, and in the case of the ria there is a simple sequence of events that creates these spectacularly wide river outlets to the sea. During the last Ice Age the sea level was lower, hence rivers that were active (obviously anything too far north would have been beneath ice) were draining down to a lower base level. As the ice melted, the sea levels rose so that river valleys were often flooded, now appearing as wide channels at the mouth of the river. Chesapeake Bay on the Atlantic coast of the USA is one of the world's largest rias, being the 'drowned valley' of the Susquehanna River.

Longest Rivers

River	Length (km / miles)	Continental area Countries drained
Nile*	6,650 / 4,135	Africa. Uganda, Tanzania, Kenya, Rwanda, Burundi, Egypt, Democratic Republic of Congo, Ethiopia, Eritrea, Sudan
Amazon*	6,400 / 3,980	South America. Peru, Brazil, Guyana, Bolivia, Venezuela, Colombia, Ecuador

Yangtze	6,300 / 3,917	Asia. People's Republic of China
Mississippi-Missouri	6,275 / 3,902	North America. USA
Yenisei-Angara-Selenga	5,539 / 3,445	Eurasia. Russian Federation, Mongolia
Huang He (Yellow River)	5,464 / 3,398	Asia. People's Republic of China
Ob-Irtysh	5,410 / 3,364	Eurasia. Russian Federation, Republic of Kazakhstan, People's Republic of China, Mongolia
Congo-Chambeshi	4,700 / 2,922	Africa. Rwanda, Burundi, Zambia, Cameroon, Tanzania, Republic of the Congo, Democratic Republic of the Congo, Angola, Central African Republic
Amur-Argun	4,444 / 2,763	Eurasia. Russia, People's Republic of China, Mongolia
Lena	4,400 / 2,736	Eurasia. Russia

* The top prize goes to the Nile, but for how long? In 2007 a new source of the Amazon was discovered in the glaciers of southern Peru (the wonderfully named Nevado Mismi peak in the Andes). The new source would make the river 6,800 km long and hence the No. 1 in this list.

HYDROLOGY

The Drainage Basin System

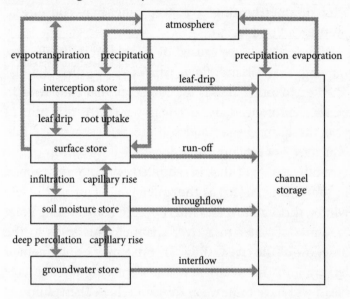

This diagram helps to show the fate of water as it passes around a drainage basin and, especially in conjunction with the hydrograph, explains floods. Overland flow is the key to flash flooding. If the rate of precipitation is greater than that of infiltration, then the water has to move over the surface. This is the fastest route for water to take to get into the channel and hence cause river discharge to rise rapidly, risking flash flooding. Thus if any conditions in the drainage basin conspire to either reduce the rate of infiltration or boost the rate of precipitation, then the flash flood is a distinct possibility. Conditions such as frozen or waterlogged ground, a hardpan on the surface after summer heating or a steep relief may conspire to produce a flood hazard.

If flash floods are caused by a large amount of water moving as overland flow, what causes other floods? The key to 'other floods', of course, is that they tend to be more prolonged and occur in winter. At this time of year the groundwater and soil moisture stores are full. You may not have put this together but the top of the groundwater (soil that is saturated with no air trapped in the soil) is known as the water table. During the late winter, throughflow and interflow are contributing great amounts of water to a river channel. This leads to the baseflow of the river rising. This rise may be augmented by, say, a storm, as any new water entering the drainage basin will have to move as overland flow. The volume of water in the river may now exceed the channel capacity and flooding may occur.

Unlike flash floods, which are sudden if not violent, prolonged flooding takes a great deal longer to recover from as the time the flood water stays high – and in your house – will be greater. Not worth dwelling on too much, but when you see those TV images of people pushed upstairs with their parrots and babies, do remember that water in their living room is not clean river water. In many countries, we have, since the nineteenth century, been easing the path of water in densely populated areas faster into rivers (preferably downstream). This means that the flood waters have invariably welled up from the sewage system. Enough said.

Deadliest river floods

Deaths	River	Date	Country
2,500,000– 3,700,000*	Huang He (Yellow River), Yangtze River, Huai River	1931	China
900,000– 2,000,000*	Huang He (Yellow River)	1887	China
500,000– 700,000*	Huang He (Yellow River)	1938	China
231,000	Banqiao Dam burst	1975	China
145,000	Yangtze River	1935	China
100,000	Hanoi and Red River	1971	North Vietnam
100,000	Yangtze River	1911	China

* The range of values indicates the difficulty in estimating the scale of losses in rural areas of the People's Republic of China. There is no doubt, however, that the 1931 flood is the greatest natural disaster in human history.

When storm surges at the coast are removed as a factor, the list above shows the extreme vulnerability of East and South East Asia to flooding. China, in particular, suffers from having large rivers sweeping across very loose loess sandy deposits. Just like a hosepipe under pressure left on the garden lawn, the rivers change course swiftly and with violent consequences. Perhaps that is at the root of the history of man's attempts to control rivers – a desire to prove ourselves through subjugating nature.

COASTS

PROCESSES AT THE COAST

As before, shout out loud, CASH! (*see* p.19) So that takes care of coastal erosion, but what about the rest of it? Well, if you've covered CASH, now it's time for a little LSD.

Longshore Drift

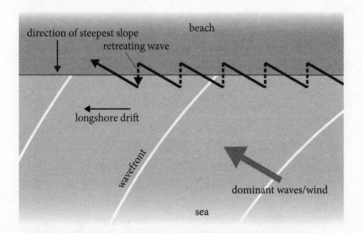

Longshore drift is the movement of sediment along a coastline. If a coastline has a dominant prevailing wind direction then waves would move as shown in the diagram.

The retreating wave, however, would move back down the beach following the gradient. Any beach material that could be moved by both the wash and the backwash would therefore end up moving along the coastline.

COASTAL LANDFORMS OF DEPOSITION

Just as the forces of erosion can carve impressive scenes with majestic landforms, so too can sediments being deposited.

The formation of a spit

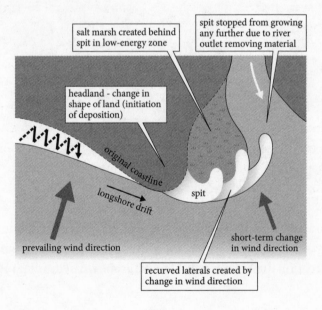

salt marsh created behind spit in low-energy zone

spit stopped from growing any further due to river outlet removing material

headland - change in shape of land (initiation of deposition)

original coastline

longshore drift

spit

prevailing wind direction

short-term change in wind direction

recurved laterals created by change in wind direction

THE SPIT

Where longshore drift takes material offshore because of a break in the shape of the coastline, a spit will naturally develop. The big guns of the spit world are Washington State's Dungeness Spit and Fairwell Spit in New Zealand. Nothing, however, gets close to the Arabat Spit in Ukraine, the world's longest at 110 km.

TOMBOLO

Essentially, if a spit reaches out to an island to link it to the mainland then the landform becomes a tombolo, as seen at Mont St Michel in France or Chesil Beach in England.

BAR BEACH

A bar occurs when longshore drift leaves a beach connecting two headlands. As with a great deal of geomorphology there are many synonyms for bars, with 'baymouth bar' appearing to be the most used, from Australia to the USA.

COASTAL LANDFORMS OF EROSION

The most majestic coastal scenery is often dominated by the features resulting from the interplay of the forces of erosion and the resistance of rock.

THE HOLY QUINARY OF COASTAL EROSION

* Headland – wave refraction creates cracks on the sides.
* Cave – the cracks get bigger, thanks to CASH.
* Arch – when the cracks meet you get an arch, and gradually the size of the arch gets bigger through constant CASH until it collapses.
* Stack – further erosion may make the stack fall to leave a . . .
* Stump – exactly that, just a stump that may be visible only at low tide.

Coastal erosion

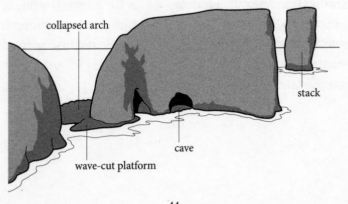

collapsed arch

stack

cave

wave-cut platform

Deepest places on the land by continent

Continent	Place	Depth (m below sea level)
Asia	Dead Sea shore, Israel-Jordan	418
Africa	Lake Vassal, Djibouti	156
South America	Laguna del Carbón, Argentina	105
North America	Death Valley, USA	86
Antarctica	Deep Lake, Vestfold Hills	50
Europe	Caspian Sea shore, Russia	28
Oceania	Lake Eyre, South Australia	15

COASTAL MANAGEMENT

Hard and soft engineering: the two ways to manage everything. These are the two contrasting methodologies and over the years, as hard engineering has been shown to be not the complete answer, gradually governments around the world have realized that instead of control structures there are times when it's better to just let nature take her course.

SOFT ENGINEERING

Managed Retreat – I just love reading those words, with their sense of calm stewardship of the man/environment relationship. Whereas it should be titled 'Run away, all hope is lost'. OK, that may sound a tad harsh but in many cases letting the sea flood where it used to is not such a bad thing for the environment.

HARD ENGINEERING

Sea wall – you spend stacks of money building a sea wall only to see it washed away, as happened at Corton near Lowestoft in October 2009. Good as sea walls are, by reflecting the energy of the waves they sometimes create new problems. Early designs often led to them actually directing the force of the waves into undermining them. And you thought it was just rivers that could think for themselves.

Revetments – a little less in-your-face than a smooth concrete wall, and a slightly cheaper option. Revetments do not aim to reflect the wave power but to dissipate it over the rocks that form their bulk, held in place by a lattice of wood that has gaps to allow both water and energy to enter. They are mainly criticized for being unsightly.

Gabions – just large steel-mesh boxes of rocks that aim

to dissipate the mighty forces of CASH. Relatively cheap, these are often the preferred choice when cost–benefit analysis has been done by the accountants.

Groynes – these stick out to sea and trap longshore drifting material to keep beaches in place. Simple it may be, but of course one beach's stolen sand is another's closer brush with erosion. The beach, is after all, a small system within the coastal conveyor belt of sediment.

SOFT & HARD ENGINEERING

Beach nourishment – here's a good idea, dredging sand from offshore and putting it on the beach you need for protecting your land. One of the cheapest solutions that does not look too unsightly. However, if the beach is for tourists, then beach nourishment may be one of the last resorts.

Offshore bars – used to reduce the impact of strong CASH action, mounds of sand are created and submerged beyond a shoreline. These features occur naturally and they are particularly good at stopping the rip current that helps create great surf waves.

TECTONICS

EARTH'S STRUCTURE

Here's another mantra to learn, repeat after me: 'Solid-Liquid-Plastic-Rigid'. Rather than Jules Verne's view of our planet being hollow and filled with dinosaurs, seismologists currently think this is what the inside of Earth looks like. Geologists have pieced it together by measuring the behaviour of seismic waves as they pass through Earth and by studying the make-up of magma, lava and escaping gases.

Of course, the most important part for living creatures is the crust and all that takes place on it, above it and within it. The problems occur when the mantle starts to move the crust.

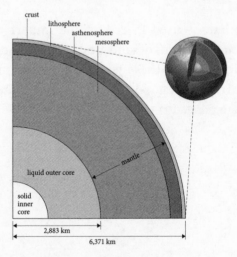

DIGGING DOWN

Jules Verne thought it possible, and during the Cold
War the US and the USSR competed to dig down into
Earth. The Holy Grail was to get to the Mohorovičić
discontinuity ('Moho') – the place where rigid crust
meets plastic mantle, which was first recognized by
the Croatian meteorologist and seismologist Andrija
Mohorovičić (1857–1936) in 1909. To date, the USSR has
got closest to the mantle, digging down during the 1970s
and 1980s to 12,261 m with what is still the deepest
hole ever dug. The American project was cancelled well
before they got to this depth. The Russians may have lost
the space race but they certainly won the depth race.

CRUSTAL PLATES

Before plate tectonics there were numerous theories about
the way Earth was structured, from Shrinking Apples to
the Hollow Earth. Seismology, palaeomagnetic survey,
deep-sea trench exploration and satellite imagery have all
combined to bring us to where we are now – Plate Tectonic
theory. But it is just that, a theory that even now is being
developed and challenged.

ALFRED WEGENER AND CONTINENTAL DRIFT

Wegener (1880–1930) was a German astronomer and climatologist who first put forward a coherent set of theories explaining all of the evidence that the continents had not always been where they are today (or, at least, back in 1912 when he first proposed his theory). From the start of the nineteenth century, more and more information was being gathered about the world during this age of exploration and a lot of the information was pored over by academics in libraries. The clearest evidence of the restless continents came from:

Mesosaurus – freshwater reptiles whose skeletons were found on either side of the Atlantic – unless they had been dropped there by a passing pterodactyl they must have once been neighbours. They date from the Permian period, between 299 and 251 million years ago. (When, of course, there were no passing pterodactyls, as they came around 50 million years later.) Similarly, the appearances of the reptilian Lystrosaurus and Cynognathus and the plant species Glossopteris across many continents added weight to the idea of continental drift.

Continental fit – early mapping showed what is much clearer from satellites: South America and Africa seem to fit together. In 1620 Francis Bacon, writing in his magnum opus *Novum Organum*, was possibly the first scholar to note the coastal correlations in a written text. With our satellite images this seems not such a great feat, but it was impressive then.

Climatology and Geology – matching strata on either side of the Atlantic and the existence of coal seams in Antarctica clearly showed that either there had been some very severe shifts in climate patterns, or the places themselves had moved.

Wegener's neat explanation of rigid plates 'floating on a layer beneath' encountered two problems: the lack of mechanism to explain how it all worked, and the assassination of Archduke Franz Ferdinand in 1914, which sparked off the First World War. The timing delayed the spread of his views until later in the twentieth century. It took another war to help push through the theory.

During the Second World War the development of a device to detect sunken submarines led to evidence being gathered for sea-floor spreading in the Atlantic. The device was able to detect small variations in the record of Earth's magnetic field held solid in the alignment of metals in the rocks below the Atlantic Ocean. This evidence from palaeomagnetism – which would have taken more time to be uncovered if the USA had not needed to find the sunken submarines – was added to the mapping of submarine mountain ranges down the mid-Atlantic, as had first been detected by U-boats and Allied submarines. The American geologist Harry Hammond Hess was the man who, in 1962, finally seemed to have provided the evidence that won over the sceptics of Wegener's theory, proving that the youngest rocks at the bottom of the Atlantic were to be found near the middle, with symmetrical bands moving west and east of rocks the same age.

PALAEOMAGNETISM – THE KEY TO WEGENER'S ACCEPTANCE

So what is this word that no spellchecker will leave untouched? 'Old magnetism' does not sound quite so scientifically enticing. But that, in essence, is what it is – rocks became magnetized at the time of their formation, however many million years ago, and that magnetization indicates the orientation and strength of Earth's magnetic field at the time. How? As basalt is extruded from the asthenosphere (the upper layer of the mantle), into the crust, it cools. As it cools, it leaves the iron-rich particles aligned to the current gravitational field.

Now that would be the end of the story if it were not for one helpful quirk: the magnetic field of the Earth is created by the spinning of the outer core, which is rich in metals, in particular iron and nickel. Dynamo theory explains that convection currents in the outer core create this field. The field is not constant – just look at the side of an OS map and you will see the legend 'True North' and 'Magnetic North'. Although the pole wanders in a pathway over our lifetimes, it is the longer timescale 'reversals' that are more useful in this instance. There have been over 170 times in the last 76 million years when the pole has 'flipped' over, leading to compass needles (had they existed) pointing the opposite way. Recorded in the new basalt of the Atlantic basin, these reversals in the direction of orientation of iron in the basalt provide the key to understanding the pace and sequence of actions at the constructive plate margin.

In the Beginning There Was One Plate

Wegener gave his theory a starting point, postulating that in the beginning there was one supercontinent. In discussions centred on his work, the name 'Pangaea' (Greek – 'All Land') then appeared. This land mass subdivided into Laurasia and Gondwanaland. Laurasia eventually tore itself apart to produce North America, Europe and Asia. Gondwanaland produced South America, Africa, India, Australia and Antarctica. There is still much debate about what may have preceded Pangaea, and the Internet is full of animations mapping the future locations of the continents.

THE PLATES

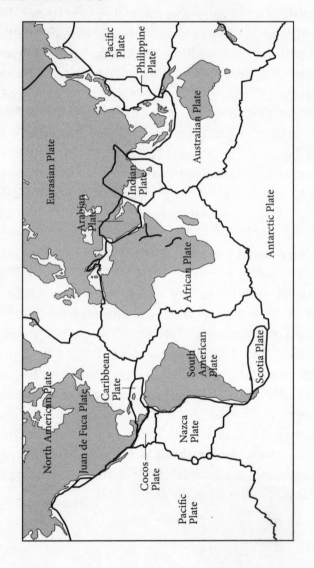

Seven large plates and various sizes of smaller plates make up the crust. Although there is a clear difference between oceanic and continental plates, it is not as simple as looking at where the seas are. Most continental plate extends beyond the shoreline out of sight, beneath the water.

Continental plate (sial)	35–70 km thick; over 1,500 million years old; density 2.6; contains silica and aluminium
Oceanic plate (sima)	6–10 km thick; less than 200 million years old; density 3.0; contains silicon and magnesium

THE BOUNDARIES

Unless the Earth was getting bigger, there had to be places where the crust was being destroyed to make way for the new crust being created. From this simple idea we get the three classic plate boundaries:

Boundary	Activity	Features
Constructive	Magma rising from the mantle creates new crust	Mid-ocean ridge, seamounts, volcanoes or volcanic islands
Destructive	Heavier oceanic crust is subducted under continental crust	Deep sea trenches, island arcs, fold mountains, volcanoes
Conservative	Plates rub past each other – may be going in the same direction but at different speeds	No new landforms except local faulting and folding

CONSTRUCTIVE PLATE BOUNDARIES

North America is moving away from Europe at a rate of 5 cm per year. The reason can be found in the middle of the Atlantic. Though the eastern Pacific boundary is longer, the mid-Atlantic constructive plate boundary is the most studied. With a buckling-up of the ocean floor, magma rising and cooling to form great seamount ranges and the occasional islands, the landform features are quite clear. Such a relatively fast rate of movement inevitably means there are earthquakes and, on occasion, surface-breaking volcanism, associated with these boundaries. Iceland exhibits the classic features created by constructive plate boundaries: tall volcanic peaks that have emerged from the sea to change from seamounts to islands. The most recent island created was that of Surtsey, off the southern coast. Between 1963 and 1968 a seamount gradually got carried away and kept erupting until eventually the seas parted and a new volcanic island was formed. This does not mean that volcanic activity has moved offshore, as since then there have been frequent earthquakes and large volcanic eruptions. The most notable were that of Hekla (1970 and 1980), the 1996 eruption beneath the Vatnajökull icecap (the Grimsvötn volcano) and the 2010 eruption of Eyjafjöll with predictions that it may have been a precursor to a new Hekla event.

DESTRUCTIVE PLATE MARGINS

Subduction is the name, creating landforms the result. Where continental shelf meets oceanic plate there can be only one winner. The continental shelf is less dense than the oceanic plate, so it is the latter that gets pushed under the former. Rather like a neat, lightweight Ferrari hitting a Chieftain tank, you know the result. The oceanic plate breaks and buckles as it is driven under the continental shelf and already we have a by-product – shallow earthquakes.

These earthquakes can produce deadly tsunami, whose effects have become all too clear since Banda Aceh in 2004. In 1964 the Anchorage tsunami killed about 131 people, but back then tsunami were treated more as natural wonders forever inscribed on oriental crockery.

As the subducted plate heads deep into the asthenosphere, it produces an oceanic trench. Just like the hunt for the Mohorovičić discontinuity, there is a continual race to see and retrieve from the depths of some of these trenches. The deepest place is the Challenger Deep, 11,033 m below sea level in the Mariana Trench, off the coast of the Philippines. Most impressive when you consider Everest is only 8,848 m above sea level.

While being bent and heated as it goes, the subducted plate melts to form plutons. These behave like wax globules in a lava lamp, rising through the crust, especially where the forces have created faults and cracks. Eventually, these plutons will either create seamounts off the coast of the continental plate or whole island archipelagos such as that of Japan. This goes some way to explaining why they have more than their share of killer earthquakes in those islands.

The final landform of note created by destructive plate boundaries are fold mountains. The greatest range of these runs from Canada to Tierra del Fuego and into Antarctica. While this is not quite a continuous range, it demonstrates the impact subduction can have on the surface of the crust. The Andes form possibly the most significant fold mountain range on Earth, rising to a highest point of 6,962 m above sea level (Aconcagua), and having a length of over 7,000 km. If you want to impress your friends you can point out that the furthest point on Earth's crust from its centre is Mount Chimborazo in the Ecuadorean range of the Andes (6,384.4 km).

On occasions the heavily folded and faulted crust of fold mountains may allow plutons to rise and dot the range with volcanic peaks. Notable fold-mountain volcanoes include Mount St Helens in the Cascade range of North America and Cotopaxi in Ecuador. Those old enough to remember the 1980 eruption of Mount St Helens may recall the violence of this type of volcano – a characteristic of destructive plate margins to be examined later.

COLLISION MOUNTAINS

There is one other different way in which destructive boundaries have shaped Earth. The Himalayas were built by the massive Eurasian plate and the Indo-Australian plate colliding. The resultant mountains can, of course, be attributed to the fact that neither of the crusts is less dense than the other. The result is a head-on collision where both plates crumple. The slightly smaller Indo-Australian plate forms the mountain root that is believed to extend as deep into the mantle as the peak rises. This is an active zone where many earthquakes provide endless case studies for teachers and students alike. Mount Everest is now thought to be growing at a rate of 2.5 cm per year, although some surveys suggest stagnation in growth.

TRANSFORM BOUNDARIES

After all the drama of Constructive and Destructive creation of new landforms, it is nice to turn to the lowly transform fault. No big landscape features, and it's very rare to have any volcanicity. The big news here, of course, is not the low-level rift or buckling of the landscape, but the earthquakes. Easily the most studied and filmed transform fault area is in California. The infamous San Andreas fault is just one of many that mark the boundary between the North American and the Pacific plates.

California south of San Francisco is in fact not on the North American Plate. The two plates are locally moving in the same direction but at different rates: the Pacific Plate is moving north-west at an average of 6 cm per year and the North American Plate is moving north-west at only about 1 cm per year. So there is a relative distancing going on on the surface of more than 4 cm per year.

EARTHQUAKES

First, an attempt to distil the greatest earthquakes in history by the number of fatalities.

Date	Location	Fatalities	Magnitude (on Richter Scale)
23 January 1556	Shansi, China	830,000	± 8
27 July 1976	Tangshan, China	250,000 – 650,000	7.5
12 January 2010	Haiti	233,000	7
9 August 1138	Aleppo, Syria	230,000	No measure
26 December 2004	100 km west of Sumatra	220,866	9.0 – 9.2
22 May 1927	Tsinghai, China	200,000	7.9
22 December 856	Damghan, Iran	200,000	No measure
16 December 1920	Gansu, China	200,000	7.8
23 March 893	Ardabil, Iran	150,000	No measure
1 September 1923	Kwanto, Japan	143,000	7.9

WHAT IS AN EARTHQUAKE?

Remember the First Law of Thermodynamics: 'Energy can neither be destroyed or created'? This is the key to earthquakes – the release of stored elastic energy in the ground. The convection currents stirring the asthenosphere transfer their kinetic energy to the crust above, rather as a rotating tyre drives a swing-boat ride at a theme park. But the crust has mass, and inertia that builds up friction between the plates. As the kinetic energy from the asthenosphere has to go somewhere, it is stored in the crust as elastic energy. It may be a large conceptual leap, but think of those elastic-band-powered aeroplanes you might have played with as a child.

Eventually, the friction-halting movement is overcome by the stored energy and there is movement. Seismic waves radiate out from the focus and are felt first at the surface at the epicentre. It is here that the greatest energy is released but, of course, the level of damage is heavily dependent on whether or not the area is built-up and, if so, on the preparedness of the humans living there.

SEISMIC WAVES

An earthquake releases energy in the form of two types of wave – Primary (P) and Secondary (S). These are longitudinal and transverse waves, and they travel at different speeds and behave differently when moving through the Moho discontinuity. Seismographs measure both types of wave and from the time difference of their arrival it is possible to calculate how far the seismograph is from the focus. All you need are three seismographs from anywhere in the world and you can locate the epicentre of any earthquake as there will be one sole overlap.

Using three seismic surveys to locate an epicentre

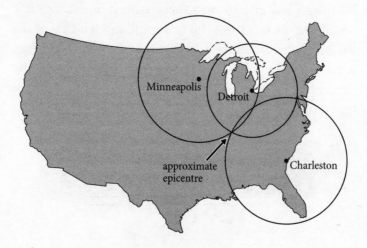

But that was only part of the theory of seismic waves. They also create surface waves and it is these that do the damage to buildings, roads and utilities. These surface waves are named Love (after the English mathematician Augustus Edward Hough Love, 1863–1940) and Rayleigh (after the physicist John William Strutt, 3rd Baron Rayleigh, 1842–1919) and it is they that oscillate the ground – either up and down or from side to side.

Seismic waves

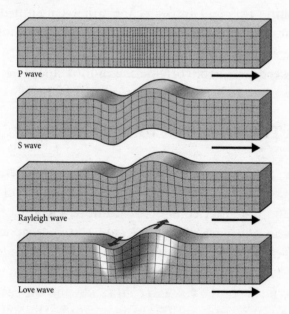

P wave

S wave

Rayleigh wave

Love wave

Tsunami

Along with landslides, tsunami are the most significant secondary natural killers triggered by earthquakes. They can result from underwater earthquake, underwater volcanic eruption, underwater landslide and landslide into water. Tsunami caused by landslides into water and landslides below the water do not scare many people – yet.

Back in 1964 the Anchorage earthquake killed around 131 people, 106 of them as a result of five tsunami. This still ranks as the worst tsunami death toll in mainland US history, but there could be more to come. Depending on whose view you believe, there may be a deadly tsunami, originating in the Canary Islands, just waiting to hit the eastern seaboard of the US.

The Threat from Cumbre Vieja

There are two mega catastrophes that threaten the US more than others: the Yellowstone supervolcano and the potential collapse of Cumbre Vieja. Cumbre Vieja is one part of the volcanic system that forms the land mass of La Palma in the Canary Islands. In 1999 it was proposed that in a future eruption the western side of the volcano could slide into the sea and create a large tsunami. This tsunami, caused by 500 km^3

of rubble sliding into the sea, would race across the Atlantic, reaching the eastern US coast in six hours. The wave is predicted to be between 30 and 60 m high at that point, and may reach as far as 25 km inland.

While several research groups agree with this scenario, there are those that do not. The main strand of disagreement is that although there is acknowledgement that Cumbre Vieja is moving, this movement is relatively slow and therefore will not create a high-magnitude-low-frequency event.

LIQUEFACTION

The terrifying prospect of your home being sucked into the ground may not have featured strongly in your teenage dreams, but if you lived in Kobe or San Francisco, it should have. When an earthquake reaches soft ground, such as reclaimed land in the Kobe port area or in San Francisco Bay, it shakes the ground – imagine shaking a bowl of sugar to level out the surface. Unfortunately, any heavy buildings sitting on the land may sink in. As simple as that – and as deadly. If you add in the displaced groundwater to lubricate the subsiding buildings, you can see that you have a recipe for disaster.

CAN WE PREDICT EARTHQUAKES?

You may have heard tales of dogs, rats and snakes being able to sense an earthquake before the foreshocks arrive but is that all we have to help us? In fact, do they sense them anyway? By using arrays of sensitive laser levels, a large number of recording seismometers and gas detectors, one could see any early signs – but by then it may be too late. The animal behaviour is now believed to be triggered by the release of gases such as radon that were trapped in the ground.

THE RICHTER SCALE

Of the two scales used to measure and compare earthquakes, the Richter scale remains the one most people are familiar with.

In 1935, while they were working at the California Institute of Technology, seismologists Charles Richter and Beno Gutenberg (after whom the boundary between the core and mantle of the Earth – the Gutenberg discontinuity – is named) developed what came to be known as the Richter scale by measuring the actual displacement of the Earth as recorded on seismometers. It is a logarithmic scale and hence ever-increasing numbers on the scale represent increases in geometric rates, not arithmetic from those below. The Richter scale has no upper limit but only four earthquakes equal to or greater than magnitude 9 have ever been recorded (or estimated):

Richter scale	Event
9.0	1755, Lisbon, Portugal
9.1	2004, Indian Ocean (Banda Aceh)
9.2	1964, Anchorage, USA
9.5	1960, Valdivia, Chile

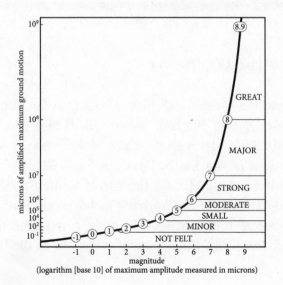

(logarithm [base 10] of maximum amplitude measured in microns)

OFF THE SCALE

It's estimated that the meteor strike that created the Chicxulub Crater off the Yucatán Peninsula of Mexico would have registered a 13.0 if the Richter scale had existed 65 million years ago. Although this is still the scale most commonly quoted in the media, the contemporary scale used by seismologists is the Moment Magnitude Scale (MMS). Developed in the 1970s, it uses the pure physical dimensions of the size of plate movement to place the earthquake on a similar scale to that of the Richter.

THE MERCALLI SCALE

Whereas the Richter scale gives a measure of earthquake magnitude, the Mercalli (also called the 'Modified Mercalli Scale') gives an indicator of the intensity of the movement of the Earth. Developed in 1902, the scale ranges from I to XII and the use of Roman numerals helped to distinguish it from the Richter scale. Owing to the inherent subjectivity of the scale, it has been much altered. Ironically, Charles Richter was one of the last to add to the revisions and to help produce the Modified Mercalli Scale that is now in place.

Highest Places on the Continents		
Continent	**Place**	**Altitude** (m)
Asia	Mount Everest, Nepal	8,848.0
South America	Aconcagua, Argentina	6,962.0
North America	Denali	6,149.0
Africa	Kilimanjaro, Tanzania	5,891.8
Europe	Mount Elbrus, Russia	5,642.0
Antarctica	Vinson Massif	4,892.0
Oceania	Carstensz Pyramid, New Guinea	4,884.0

VOLCANOES

By far the most impressive tectonic events to capture the imagination, be it in Dante's descriptions of hell or Pliny's description of Vesuvius erupting, volcanoes hold a special fascination for us all.

TYPES OF VOLCANO

There are five different types of volcano:

Fissure – not your classic symmetrical cone, these are large cracks in the crust through which lava emanates.

Though low down in the fame stakes, of all the volcanoes these have been the single largest contributors to global climate change and large-scale landscaping. The Deccan Traps in central India illustrate the scale at which these features can operate. Here there are multiple layers of basalt, 2,000 m thick, covering an area of 500,000 km^2.

Shield (basic) – shallow-sided and broad, shield volcanoes are caused by relatively pure basalt running down from the summit rapidly while cooling. These are typical of constructive plate margins and hot spots. Hence the Hawaiian Islands and, to a degree, Iceland are covered in these features.

Dome (acidic) – steep-sided features such as Mount Pelée on Martinique. These are the 'classic' volcanoes and are mainly produced at destructive plate boundaries. With a high level of silica mixed in with the basalt, the lava of these volcanoes is very viscous and solidifies before running too far down the slope. In addition, these volcanoes tend to erupt violently as the viscous lava will trap steam and gas within it until it can hold it back no more. As with a shaken bottle of champagne, when the pressure builds too much, the top will blow off.

Composite – (also called a strato volcano) creates layers of lava and ash/cinders as it goes through a long sequence of eruptions. The result is a composite with layers of hard lava and less consolidated ash from violent eruptions. This layering produces weaknesses that can be exploited by the magma and produce eruptions with large amounts of ash in their ejecta. This is what the inhabitants of Sicily can expect from Mount Etna.

Calderas – the volcano with the lake in the middle. When the eruption has been large enough to empty the magma chamber, the ground will slump down to fill that void. The result is a depression in the summit of the volcano and subsequent infilling with a lake. While this is a fairly common feature to be found from Africa to North America, the most famous example must be Santorini in the Mediterranean. Here, the sea has broken through the walls of the caldera and produced a rather impressive ring of small islands.

WHAT'S THE DIFFERENCE BETWEEN LAVA AND MAGMA?

Both molten rock, but while magma is molten rock below the Earth's surface, lava is magma that has reached the Earth's surface.

WHAT COMES OUT OF A VOLCANO?

Each volcano is different but when they erupt they eject four broad categories of matter:

Lava – spectacular on video, but not too dangerous. There are two broad divisions of lava and their behaviour is closely linked to the level of silica in the magma.

> **Basaltic** – low-silica, free-running, found at constructive boundaries with frequent small eruptions, hence easy to spot where the problem is and not get in the way. There are many local names for this type of lava, a favourite being 'Aa' – worth trying in Scrabble – a free-flowing lava of the Hawaiian Island chain.
>
> **Andesitic** – the opposite of basaltic, full of silica, very viscous, very scary. A cooler, slower-moving lava that is usually found near destructive plate boundaries. The extra ingredients added during subduction are the sediments on the seabed, rich in silica. The result is that when cooked up, impurities in the form of gases are released under pressure, making for a violent and less frequent set of eruptions. However, if eruptions are less frequent people downgrade the risk – at their peril.

Pyroclastic flow – a rather general term for hot material sent out from a volcano. This covers all manner of things that travel down the sides of mountains. The traditional image is of the burning cloud or *nuée ardente* moving and cremating all in its wake, as at Pompeii or from Mount Pelée. *Nuées ardentes* have variously been ascribed the following range of characteristics: a top speed of 300 m per second, with a temperature as high as 1,200 °C. In addition, airborne fragments ejected by an eruption ranging in size from huge lava bombs through smaller stones, to ash, are dropped on the volcano sides during an eruption. Hence there will be a mix of rock fragments ready to be taken up by any passing mudflow (lahar) as it flows down the volcano's slopes.

Ash – very fine airborne fragments that have a local effect in the make-up of lahars, a regional effect in the devastating of crop lands and a global effect in altering climates. We may need to revisit that comfortable idea of volcanoes creating fertile land – it is not easy to tell an Aeta person, displaced by lahars and ash from Mount Pinatubo in the Philippines (1991), that it will be OK once the soil has started to develop.

Gas – as you can see from the section on climate change (*see* p. 97), the gases emitted by volcanoes are not to be ignored. On a local scale they provide a good indicator of whether an eruption is imminent and can be seen as good news. To the people living around Lake Nyos in the Cameroon in August 1986 the gas proved less useful. A crater lake into which magma leaks carbon dioxide from beneath, Lake

Nyos suddenly emitted a giant noxious cloud of carbon dioxide mixed with hydrogen sulphide, which spread over nearby villages, killing some 2,000 people and destroying their livestock.

NEVADO DEL RUIZ

Nevado del Ruiz is the northernmost volcano of the Andean Volcanic Belt, lying about 130 km west of Bogotá in Colombia and within the Pacific Ring of Fire. It has been active for about two million years. In November 1985 a relatively small eruption produced pyroclastic flow, which combined with melted snow and ice from the peak and created several large mudflows, which in turn flowed down the volcano's sides at speeds up to 60 km per hour, destroying all before them and growing as they went – one reached a width of 50 m. One of the mudflows almost entirely engulfed the small town of Armero, the others were nearly as devastating. Altogether over 5,000 homes were destroyed and more than 23,000 people killed in what came to be known as the 'Armero Tragedy'.

ERUPTION TYPES

There are two types of volcanic eruption – rare, volatile and downright dangerous or frequent, spectacular but 'mostly harmless' (as Douglas Adams put it). Within these two bands we group all eruptions on the land:

Mostly harmless – Icelandic and Hawaiian eruptions are so frequent and relatively gentle that unless they have unforeseen secondary impacts (as with the melting of the Vatnajökull ice cap after the eruption of Grimsvötn in October 1996), they can be considered the least dangerous. At the top of this group are the Strombolian eruptions that, though less frequent and more violent, are again easily predicted, giving enough time to run away.
Downright dangerous – The names say it all –Vesuvius, Krakatoa, Mount Pelée – three of the most famous and deadliest of volcanoes.

MEASURING VOLCANIC ERUPTIONS

Volcanoes' eruptions are measured using a logarithmic scale called the Volcano Explosivity Index (VEI), with a maximum value of 8.

The VEI was developed in 1982 by the volcanologist Chris Newhall at the United States Geological Survey

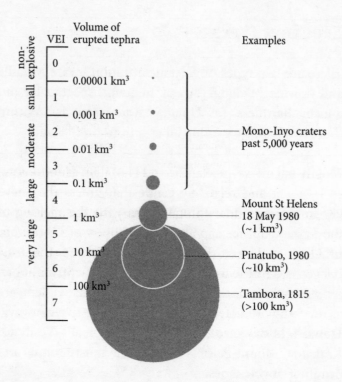

(USGS) and Professor Stephen Self of the University of Hawaii. It brings together quantitative and qualitative measures: volume of ejecta, height of the eruption cloud and duration of the eruption. The adjectives used to describe each classification make for great reading and, much as the Mercalli scale adds another dimension to the Richter scale, so these adjectives 'enliven' the VEI:

VEI	Classification	Description
8	Ultra-Plinian	Mega-colossal
7	Ultra-Plinian	Super-colossal
6	Plinian	Colossal
5	Plinian	Paroxymal
4	Pelean	Cataclysmic
3	Vulcanian	Severe
2	Strombolian	Explosive
1	Hawaiian	Gentle

Plinian refers to the two Plinys: Pliny the Younger wrote about the eruption of Mount Vesuvius in AD 79, which killed his uncle, Pliny the Elder. Pelean refers to Mount Pelée on Martinique; the best example of this scale of eruption was Mount St Helens in 1980.

SUPERVOLCANOES

Bringing things up to date on the subject of volcanoes, it would be remiss of me not to mention 'supervolcanoes', one of which lies beneath Yellowstone National Park in the US. The evidence from Yellowstone is that there is a very large magma chamber probably filling up under the park, and that a huge volcanic event there is long overdue. Evidence of previous eruptions tells us the scale and chronology of these events. In terms of chronology we are within an eruption time period and in terms of scale there will almost certainly be a catastrophic physical impact in the USA and a climatic impact around the world.

WHY DO PEOPLE LIVE NEAR VOLCANOES?

There are some places where the volcanic hazard is minimal and the advantages of the area outweigh the risk. Iceland is a classic example – all that geothermal power for free and so much hot water that it is piped centrally into homes and, after being used for heating, flows out under the adjacent roads to keep them snow free. For the Aeta on the slopes of Pinatubo there are various benefits (once the land has stabilized), from flat land rich in nutrients to making pottery. For each location there will be potential costs in the future and benefits in the present. With the aid of better prediction and reduced vulnerability afforded by engineering, especially in wealthy countries, the cost–benefit analysis swings in favour of current benefits. Of course, if you are able to get insurance for your home or business, this is made all the more acceptable. But in Less Economically Developed Countries (LEDCs) the truth is starker: there is not much choice.

CLIMATE
AND WEATHER

Always a tough one, the concepts were too abstract. All too often, despite it being around us all the time, we just found it difficult to explain why it was always so cold and wet. Before delving into some of the key aspects of this section, here is a mantra you must learn:

THE MANTRA OF METEOROLOGY

* Warm air rises
* As it rises it cools
* As it cools, its ability to hold water as vapour (the gaseous state of water, of course) reduces
* If relative humidity reaches 100 per cent, condensation will occur
* Clouds form
* Precipitation may result

If you get this, then what follows will be child's play.

What Causes Rainfall?

One of the easiest questions to answer: 'air rising'. It really is that simple:

Relief (Orographic) rainfall – air blows towards a mountain; it cannot blow through it so it is forced to rise. Go to stage 2 of the Mantra of Meteorology.
Convectional rainfall – air heats up by contact with warm ground. It is less dense so it rises. Go to stage 2 of the Mantra.
Frontal rainfall – at a warm front, warm air rises over cooler air. At a cold front, the cold air forces warmer air upwards. In both cases, go to stage 2 of the Mantra.

General Atmospheric Circulation

To start at the largest scale – and really where it does all start – the climates of the Earth exist to move energy from where it is abundant to where it is in deficit. That means, of course, from the equator to the poles. If the equator is a line that joins up points on Earth's surface that are equidistant from the two poles, then the thermal equator is the line of latitude that gets the most insolation – exposure to the sun's rays – on any one day.

The Tri-cellular model or General Atmospheric Circulation model

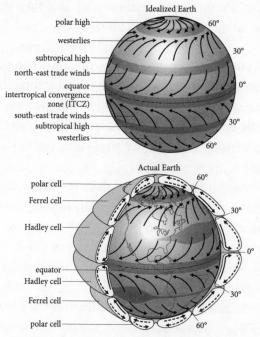

At the thermal equator, insolation heats the ground, the air in contact with the ground (the boundary layer) heats up, becomes less dense than the air around it and thus rises. Ready for the mantra? 'As the warm air rises' etc. Hence above the thermal equator you will find a band of rain – the equatorial rainbelt – which explains the location of the lushest biome in the world, the tropical rainforest biome.

The origins of the tri-cellular model can be traced to George Hadley (1685–1768), who first put forward the idea, in 1735, that equatorial air rises and then diverges. In 1856, William Ferrel (1817–91) developed Hadley's theory, and hence it became enshrined. As most of our climatic interactions take place in the lowest 10 to 15 km of the atmosphere (troposphere), the air that has risen must move north or south when it reaches that height. An analogy would be a lid trapping rising air. The air cools as it rises and so, having moved north or south, it will descend. (Excuse me if I seem to ignore the southern hemisphere: I have learned that it just complicates things to try always to refer to both hemispheres.) The air descends approximately 30° north of where it rose. The air descending creates Sub-Tropical High Pressure (STHP). At this point you can guess the impact on the local region: the descending, warming air leads to dry conditions and the existence of great arid zones like the Sahara Desert. Once descended, the air has to diverge at the Earth's surface. Some air returns equatorwards and completes the Hadley Cell, while some heads north. The air that moves north meets air from the polar cell that has been super-cooled over the Arctic and thus, being dense, has descended and been deflected south.

Now, it just so happens that if you are reading this tome in the UK, then the chances are that this convergence is happening above your head. The UK is blessed with its joyous weather because of the beautifully named Polar Front of Convergence (PFC). That's warm air of the Ferrel Cell meeting cold air of the Polar Cell. Much as with the continental plate meeting the oceanic plate, something has to give. That something is energy. The air mixes, the Ferrel air cools and, of course, when air cools the fun and games begin.

Before leaving this macro view it is worth pointing out two things: firstly, that ocean currents do a lot of energy transfer too; and secondly, that the thermal equator moves with the seasons. As a result of the thermal equator moving, the equatorial rainbelt moves and the PFC moves. Thus you get the Okavango swamps, for example, fed by summer rains that then dry up in the winter.

THE DEPRESSION

Correctly put, this should be called a temperate cyclone, and it shares some characteristics with tropical cyclones – with one clear difference. To produce a depression, take some contrasting air masses and add a twist. The PFC gives us our contrasting air and the twist comes from a centre of low pressure. Air pressure, be it high or low, is simply a measure of the weight of the air pushing down on the Earth's surface. In deserts the air is descending and hence you have high pressure, whereas when air rises it relieves the pressure of the air, leaving low pressure at the surface. The formation of a sequence of high- and low-pressure cells under the PFC owes itself to the jet stream.

THE JET STREAM AND THE LOW-PRESSURE CELL

The Northern Hemisphere Polar jet stream

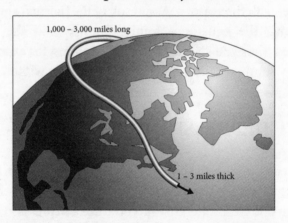

There are few occasions when we have to look higher, past the troposphere, to get an explanation of our weather, but this is one of them. Jet streams blow above the layer of atmosphere that contains our weather. They are found in the tropopause (10+ km above sea level) and blow incredibly quickly, up to nearly 40 km per hour.

Famously used by jet airliners, the jet stream causes a succession of high- and low-pressure centres at the latitude of the UK. The jet stream blows in a wave shape (the Rossby Wave), accelerating as it moves from north to south and vice versa, decelerating as it turns. Rather like a car sweeping down a leaf-strewn road in a Hollywood film, sucking leaves up behind it as it goes, the accelerating jet stream sucks air up from the troposphere and produces a low-pressure cell to mess up our weather. Of course, as the air in the jet stream slows down it has the opposite effect: air piles up and is forced down to give a high. This is why you often see a sequence of high- and low-pressure cells coming across the Atlantic.

THE DEPRESSION AND ITS WEATHER

A classic depression

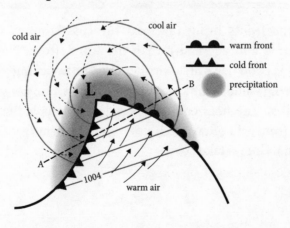

I expect this image is familiar. The depression is made up of the classic warm and cold sectors plus the cold and warm fronts. First, notice that the winds are blowing inwards and anticlockwise (this is known as the Coriolis force, named after a nineteenth-century French scientist). Low pressure means anticlockwise winds; high pressure or anticyclone gives outward and clockwise winds.

Cross-section through a depression

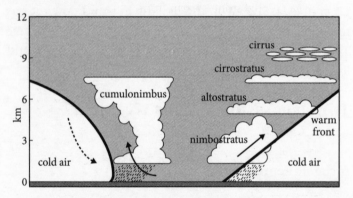

So, the fronts bring rain and the cold front is more dynamic and aggressive than the warm front. Neither front is good news on a weather map as they bring lots of air rising – and I think we've covered what happens when air rises. The final comment on depressions is that the cold front will often 'catch up' with the warm front, and when it does – take cover.

An Occluded Front

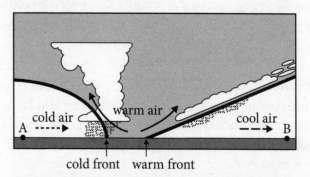

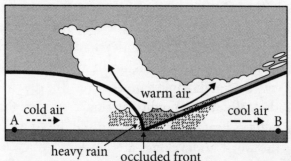

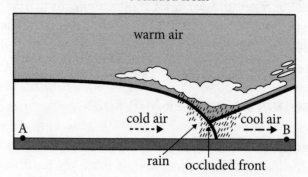

Great – now not only is the warm and more moist air being forced upwards, it is totally surrounded by cold air. The result is high winds, dark clouds and plenty of rain.

THE ANTICYCLONE

By contrast with that dastardly depression, the anticyclone is a cinch. Air descends in a high-pressure cell. Winds are relatively calm; there is no coming-together of air masses. The weather can be rather nice, especially if there is a tropical air mass involved. The seasonal extremes do cause havoc. In the summer anticyclones stop any cloud forming and can lead to droughts. In the winter they can bring very cold night-time temperatures. The descending air leads to limited amounts of cloud, so at night the Earth radiates heat, none is reflected and the air gets very cold. The worst anticyclones are blocking anticyclones. These stay around for a lot longer than normal as the jet stream can split and hold them in place, rather as an eddy current in a river will stay in the same place, while water passes through it.

THE INDIAN SUMMER MONSOON

Cherrapunji, also known as Sohra, (Meghalaya, India) prides itself on being the wettest place on Earth, with a record year of 22,987 mm of rainfall between August 1860 and July 1861. When you consider that 1 mm of rain equates to 1 litre of water per square metre, you might begin to see the problem. Nearby, Mawsynram has been claiming the title over recent years but weather-nerds are loath to pass on the crown. This resistance is due to the limits of the continuous record of precipitation there. The region in which these two places are found experiences the Indian summer monsoon, which is a feature climate of the globe and a staple of geography knowledge for all.

Monsoon Winds

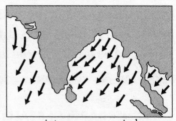

summer monsoon winds   winter monsoon winds

THE THEORY OF CONTINENTALITY

Put simply, the centres of large continents get hotter in the summer and colder in the winter than coastal places on the same latitude. Whilst the seas move heat readily around their great basins, land has a lower specific heat capacity. Which simply means that less insolation is needed to raise the temperature of land than water. Which means it gets hot rapidly in the summer. Over Siberia, north of the Tibetan plateau, is the largest area where this phenomenon occurs. During summer the Siberian interior gets heated and air in contact with the land warms up. This less dense air rises, causing a regional-scale low-pressure cell, which is so large and so deep that air from over the Arabian Sea and the Bay of Bengal moves on to the Indian subcontinent. The air, of course, is very moist and hence it brings rain across India. Over either the Western or the Eastern Ghats the air is also forced to rise and the rainfall is intense. An extra-special ingredient for the Cherrapunji region is that it nestles in the foothills of the Himalayas. Yes, you guessed it. The wet air is drawn up this great mountain range and as it rises . . . well, it rains. It rains a lot.

IS CHERRAPUNJI REALLY THE 'WETTEST PLACE ON EARTH'?

Well, no. In the winter, as any lovers of epic Russian literature will know, Siberia gets very, very cold (the record, on 6 February 1922, of -67.8 °C). Cold land means descending air, which means high pressure. The result is that in the winter everything is reversed and Cherrapunji now finds itself in the 'rain shadow' of the

Himalayas. Because in the winter months Cherrapunji does not get much precipitation, a third contender has appeared to rain on the parade (excuse the pun) of the other two: Mount Wai'ale'ale, on the island of Kaua, Hawaii, is most widely accepted as the wettest place on Earth now, as the data is deemed to be more reliable and it can have up to 360 days of rain per year.

HURRICANES, TYPHOONS, CYCLONES AND THE WILLY-WILLIES

This is the stuff of legend and the cause of plenty of human misery – Tropical Revolving Storms (TRS). Tropical Revolving Storms are regionally named, so you will not get a typhoon in the Caribbean or a hurricane over Bangladesh. If you were wondering, you'll find willy-willies in Australia.

TRS FORMATION

It is rare for a TRS to develop if not all of the conditions below are met (though similar storms have occurred in the Mediterranean as recently as September 2006). To create TRS you need:

* Water of 26.5 °C to a depth of 50 m.
* A Coriolis force (*see* p.86) – thus no closer than 500 km from the equator, where there is negligible Coriolis effect and you do want this thing to spin.
* A storm convection cell – to give that low-pressure centre.
* Negligible shear wind – wind blowing horizontal to the sea level – it stops the convection towers from brewing up.

How a TRS develops

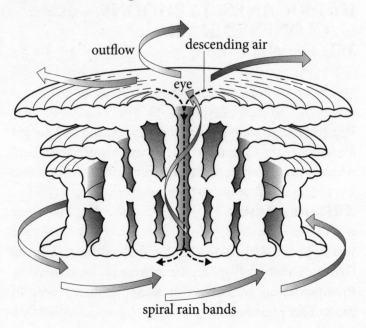

outflow   descending air

eye

spiral rain bands

Water of the heat specified above evaporates readily and takes heat energy into the system as kinetic energy. Towering convection clouds take the rising air to great height very rapidly and the water vapour condenses. If one has to add energy to a kettle to create steam, it also follows that, when the opposite effect occurs, condensation, then energy is coming out of the rising air and that is the source of the great kinetic energy of the winds. The low pressure sucks the air inwards and the Coriolis force makes the air spiral.

WHAT DOES THE DAMAGE IN A HURRICANE?

After Hurricane Katrina (2005) and Cyclone Nargis (2008) you may have clear ideas of the impacts of tropical cyclones, but there are three agents of destruction:

High winds – windspeeds above 118 km per hour get an ordinary tropical storm on to the hurricane scale, but windspeeds have been estimated to reach in excess of 306 km per hour. These winds are enough to blast away trees, cows and buildings, but if you add in the debris they carry you have a formidable factor. (The highest ever recorded windspeed is that of 408 km per hour during Tropical Cyclone Olivia in 1996 in Barrow Islands, Australia.)
Precipitation – intense precipitation up to a record of 211 mm for Hurricane Agnes (1972) causes local flooding

and direct raindrop and hail impacts.

Storm surge – the storm surge is often cited as the most significantly destructive aspect of tropical revolving storms. They are produced by strong winds at the leading edge of a storm, coupled with the 'bulging' of the sea due to the low-pressure centre. Katrina produced a maximum storm surge of 7.6 m as it hit the Mississippi coast.

TORNADOES

Like a whirlpool in a river, tornadoes are eddies in the atmosphere. Strange as it may seem, the UK has the highest frequency of tornadoes of any country in the world; however, they are normally concentrated in the lower levels of the Enhanced Fujita Scale (0 and 1). The USA has over 1,200 tornadoes per year, many across Tornado Alley, where most are above Enhanced Fujita Scale 2.

When warm air and cold air collide, large cumulonimbus clouds will often result and it is from these that tornadoes descend until they make landfall. With windspeeds from 64 to 177 km per hour, these short-lived phenomena are highly destructive but lack the scale, rain or surge to match the devastation caused by Tropical Revolving Storms. The suddenness of their appearance means that little or no preparation can be made for them. At their most beautiful, when tornadoes pass over water they create waterspouts, sucking water high into the air. Urban legend has it that they have taken up unsuspecting divers from the sea, later depositing them offshore. Almost certainly they have taken up fish, leading to them raining down inland.

EL NIÑO

This is one of the most significant climatic features that causes extreme weather. Normally the trade winds in the Pacific blow from South America towards Asia. For some not yet fully understood reason, on occasions these winds fail. This means that the warm water that is usually piled up in the western equatorial Pacific spreads across to the east. The result is rather like turning on the second bar of your electric fire. The atmosphere gets a lot more warm sea surface to give it energy and more evaporation gives it moisture. As a result, the atmosphere springs into a different state and changes the weather.

Some impacts of El Niño

Western impacts	Eastern impacts
Drought in Indonesia, Australia and Malaysia	High rain in South America – leading to landslides along the Andes
Mild winters in northern USA and western Canada	Heavy rainfall in Europe in spring causing flooding
Reduced hurricane season	Demise of Peruvian fish catches
Increased Galapagos iguana mortality rate	Increased rain in southern USA states

Various historical annihilations of ancient civilizations have been blamed on El Niño (Spanish for 'the little boy'), including that of the Aztecs. It is also believed that the weather anomalies helped create the poor harvests in western Europe that may have contributed to the French Revolution (1789).

There is also an effect called La Niña whose impacts are the reverse of her brother. Together the occurrences are known as ENSO (El Niño Southern Oscillation). The bad news is that climate change will not only make the effects bigger but also make them more frequent.

GLOBAL ISSUES

CLIMATE CHANGE

Climate change is one of those areas where there is a great deal of evidence that is almost indisputable and then there are the theories on various scales that compete for the top spot. Science works by supporting theories, building up evidence, feeling not too happy with our theories as complete explanations, coming up with new theories, supporting those, knocking them and putting new ones forward. With respect to climate change, we do not yet understand all of the interrelationships involved in terms of timescales and sizes of changes, and apparently small changes in the amount of energy trapped in the Earth/atmosphere system can have profound knock-on effects. These effects are only now being partially understood.

In other words, there are great uncertainties in this subject and massive complexity – not all of it well enough understood to be predictable. At time of going to press there is a near consensus in two broad areas:

* Climates have always changed in the past and will change in the future (with or without humans).
* Humans have altered the natural system of cycles

such as the carbon cycle so that we may now be in a different place with regard to our climatic zones and hence our weather.

Beyond this we are still in the realms of research, theory and evidence. This includes the very recent evidence of global climate change – the relative lack of warming this century.

Two 'Knowns' of Important Understanding

Albedo and the globe – Albedo is simply the measure of the reflectivity of the Earth or ocean surface (for example, mirrors are almost 100 per cent reflective – they have a high albedo; dark green tropical forest has a low albedo of only 14 per cent). So what does that mean? Well, if you reflect less you absorb more. If a surface on Earth absorbs more, then the sun's energy is going to be trapped in our planet's environment. A great deal of the energy is re-emitted in the long wave and trapped by the greenhouse gases. Reflected light goes straight back out to space, as reflected waves do not change their wavelength and if they came through the atmosphere then they can just go back where they came from.

One last point on albedo: the northern hemisphere is 39 per cent land, whereas the southern is only 19 per cent. This means that on a global scale the seasons result in a change in the global average albedo. Hence if we alter the seasonal variations in sunlight hitting Earth, we will alter the overall annual balance of energy reflected compared with that absorbed.

The greenhouse effect – Put simply, without the greenhouse effect we could only really survive on a narrow equatorial band of land and ocean. The gases that make up the atmosphere of the Earth are good at trapping outgoing long-wave radiation. This outgoing radiation comes from the Earth as a result of it being heated up by the short-wave radiation from the sun. If the gases did not do this fiendish trick then the average temperature of the Earth would be -18 °C compared with its current +14 °C. If you consider that average global temperature needs to drop by only 6 °C to put us into the worst of all ice ages, you can see why we love the greenhouse effect.

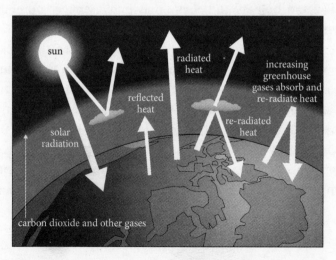

THE KNOWN KNOWNS WRAPPED UP

According to the US Environmental Protection Agency, the following items should be added to the list:

* Human activities are changing the composition of the atmosphere.
* An unequivocal warming trend has been recorded in both hemispheres between 1906 and 2005.
* Most greenhouse gases have atmospheric lives from decades to centuries and thus the levels of these gases are almost certain to rise in coming decades.
* Increasing greenhouse gas concentrations tend to warm the planet.

A LIST OF GREENHOUSE GASES

It would be easy to believe that public enemy number one, carbon dioxide, was the sole greenhouse gas, or at least the most significant. In fact, it isn't even the most powerful natural greenhouse gas. Methane can be up to twenty times more potent as a greenhouse gas than carbon dioxide. Add to that the fact that methane naturally creates carbon dioxide when it oxidizes in the air and you can see how important it is. With large amounts of (naturally occurring) methane trapped on the Earth's crust, a great deal of it poised to escape to the atmosphere through permafrost thawing, you can see the potential problems ahead. Notice we have not yet mentioned the increase in flatulent cattle that grab the headlines for the increase in methane levels.

The gases	Contribution to the greenhouse effect
Water vapour	36–73 per cent
Carbon dioxide	9–26 per cent
Methane	4–9 per cent
Nitrous oxide	6 per cent
Ozone	3–7 per cent

The large ranges of values for each gas stem from the universal vat of uncertainty that remains little reported but is the elephant in the room of this subject. In addition to the gases listed there is a whole suite of synthetic gases made and used by humans: halofluorocarbons (HCFCs) and chlorofluorocarbons (CFCs).

CHANGES IN THE CONCENTRATIONS OF GREENHOUSE GASES

Central to the work of the Intergovernmental Panel on Climate Change (IPCC) and a great many politicians, activists and entrepreneurs is the rate at which the levels of greenhouse gases have been rising.

Greenhouse gas (by concentration)	Current concentration (ppm)	Estimated annual increase parts per million per year (ppmyr-1)
Carbon dioxide (CO_2)	387	1.6
Methane (CH_4)	1.751	zero
Nitrous oxide (N_2O)	0.001 - 0.314	0.06

It is upon these small increases that science falls to make the link between man's activity and global warming.

The Known Unknowns

I would rather label these the 'nagging suspicions': we just do not know enough about this list for many people's liking, but it is having that opinion that gets you labelled a heretic. Broadly speaking, the following points are the very complex issues that face humans before we even get down to considering the next step forward:

* Improving understanding of natural climatic variations, changes in the sun's energy, land-use changes, the warming or cooling effects of pollutant aerosols, and the impacts of changing humidity and cloud cover.
* Determining the relative contribution to climate change of human activities and natural causes.
* Projecting future greenhouse emissions and how the climate system will respond within a narrow range.
* Improving understanding of the potential for rapid or abrupt climate change.

THE KEELING CURVE

Since 1958 we have access to a continuous record of the atmospheric CO_2 levels thanks to the enlightened work started by Charles David Keeling.

The Keeling Curve

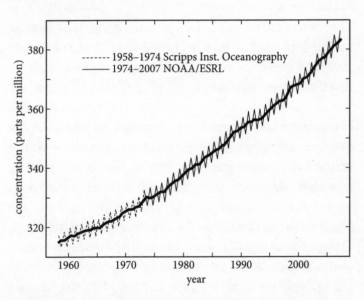

Initially, Keeling measured the carbon dioxide content of the air at the South Pole and on Mauna Loa in Hawaii, but from the 1960s only the Mauna Loa readings have been made and added to the curve. The graph shows the seasonal impact of the greater amount of vegetation in the northern hemisphere photosynthesizing and removing the gas from the air in the summer. This graph is a cornerstone in the climate-change debate as it clearly indicates rapid increases over time in the atmospheric level of carbon dioxide.

PALAEO-CLIMATES (THE PAST)

The best way to predict how something will change in the future is to look at how it has changed in the past, and there are various techniques that will help in discovering past climates.

Stratigraphic sequencing – younger layers of snow and ice, sediments in lakes and ocean depths lie on top of the old, trapping their contents. These will include minerals, plants and especially the skeletons of animals (exo and endo if you want precision). The best records come from ice cores, lake beds and abyssal plains.

Oxygen isotopes – ^{16}O and ^{18}O exist in balance according to the climate. Hence if you can extract the oxygen from the air bubbles in an ice core and measure the balance between the two, you can extrapolate the climate of the time the air was trapped. Without this technique there would be a lot fewer holes dug in Greenland and on Antarctica.

Radiometric dating – usually known as radiocarbon dating. Isotopes of the same element decay at a known rate. So if you look at the ratio of isotopes in your sample from a certain depth, you can figure out when the material was deposited. This works for ^{12,13,14}C (the three isotopes of carbon) up to 100,000BP, krypton and argon back beyond 500,000BP and uranium back to 350,000BP

Palaeomagnetism – useful for dating really large-scale changes in crustal movement (*see* Tectonics, p.48).

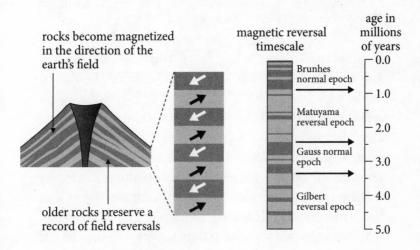

rocks become magnetized in the direction of the earth's field

older rocks preserve a record of field reversals

magnetic reversal timescale

age in millions of years

Brunhes normal epoch

Matuyama reversal epoch

Gauss normal epoch

Gilbert reversal epoch

0.0
1.0
2.0
3.0
4.0
5.0

Data from the Vostok ice core

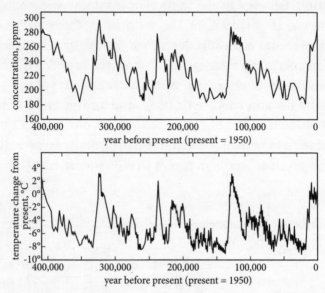

THE VOSTOK ICE CORE RECORD

First published in 1999, this graph is possibly the most famous one on this subject, and shows the result of the ice core that was drilled just over 2.2 miles down into the Antarctic ice. The long timescale, going back over 400,000 years, allows us to clearly see the way the climate of the world has oscillated since before we all drove cars, flew round the world and were racked with guilt. Clearly, there needs to be an explanation for this record of climate changes without resorting to human actions.

The Factors Influencing Climate Change

There are myriad explanations for the natural oscillation in the global climate as shown by the Vostok ice core.

It makes sense to look first at the big source of energy in the sky and how we relate to it before we look at the Earth itself. After all, if you are hot in your living room you turn the fire down; if you are cold you move closer to your fire, don't you?

The Milankovitch cycles

Milutin Milankovitch was a Serbian mathematician and civil engineer, with an interest in applied problems. He calculated that the way we relate to the sun is not constant. He found there were three cycles that characterized our relationship to the sun, and these therefore affected the amount of useful sunlight energy our planet would trap. Think of it this way: if global warming is all about the amount of energy our atmosphere stops from leaking out to space, then Milankovitch was all about the amount of energy we trap in the first place. The link between the cycles and the effect on the Earth's temperature is not simple, but it relates to the seasonal amount of energy received and that relates to the difference between the northern and southern hemisphere.

The cycles

Eccentricity (orbital shape) – every 100,000 years the Earth moves from orbiting in a circle around the Sun to going around it in an elliptical shape. You may not care, but this is due to the pull of Jupiter and Saturn (remember 'large bodies attract'). The more circular, the more continuous the energy received across the seasons and vice versa.

Axial tilt – every 41,000 years, the Earth changes the angle it lists on its spinning access, from 21.5° to 24.5° and back. This alters the angle of incidence at which the sunlight hits the Earth at different times of the year. Hence over 41,000 years there is a variation in the amount of light reflected and absorbed.

Precession – has the shortest timescale. Every 23,000 years the imaginary axle that runs from the North Pole to the South Pole draws a circular shape. Once again, there will be times in this cycle when on balance the Earth absorbs more energy, and there will be times when it absorbs less.

At present it is only precession that is in the glacial mode, with tilt and eccentricity not favourable to glaciations. These three cycles, with their own maximum and minimum effects, superimpose to give a resultant impact on the Earth. You may have switched off in your physics lessons but when waves meet they interfere and when three waves meet the result can be chaotic.

The resultant waves of the Milankovitch cycles

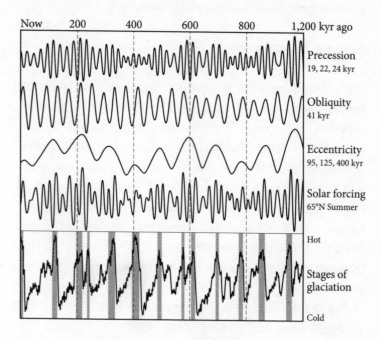

| Now | 200 | 400 | 600 | 800 | 1,200 kyr ago |

Precession
19, 22, 24 kyr

Obliquity
41 kyr

Eccentricity
95, 125, 400 kyr

Solar forcing
65°N Summer

Hot

Stages of
glaciation

Cold

MILANKOVITCH CYCLES AND GLACIAL PERIODS

The diagram above shows the three cycles and the result of their overlapping 'Solar Forcing'. You may or may not notice any correlation between the pattern of the lower two graphs. If you don't, then good for you – you are an innate believer in global warming. If, however, you do see

109

a correlation, keep it to yourself because you are showing early symptoms of being a climate heretic.

SOLAR VARIATIONS

You should by now be getting the idea that isolating one single factor as a cause of climates changing is like blaming one player for the loss of a football match. In addition, even the Sun itself turns out to be pretty unreliable. The Sun has an eleven-year cycle that leads to the energy getting to the Earth's atmosphere varying by as much as 0.1 per cent. It is agreed that this may have been a significant contributor to the Little Ice Age from 1400 to 1700, but it may be dwarfed by contemporary changes to the greenhouse gas composition.

A BRIEF RÉSUMÉ OF OTHER NATURAL EFFECTS ON CLIMATE CHANGE

Once we have dealt with the issue of the variations in energy getting into the atmosphere and being trapped, are there any other mechanisms we can isolate? The factors below are given less space merely because we had to make cuts somewhere; they may all or partly be just as significant.

Volcanoes – if the eruption of Mount Pinatubo in 1991 could cool the northern hemisphere by 0.5 °C for two years afterwards, you can see how volcanoes could play their part in changing climates. Supervolcanoes or VEI 8 'super eruptions' are believed to have been large enough to have created short-term ice ages. Evidence of the scale of these events comes from the infamous Deccan Traps basalt region of India. Volcanoes cool climates by increasing the amount of backscatter of sunlight, by putting rock and soot fragments into the air. In addition, they emit large volumes of sulphur, which alters the greenhouse effect to reduce temperatures. One of the methods proposed to reduce current global warming is to add more sulphur to the atmosphere in various ways, which seems a very odd suggestion to me.

Continental drift – as you stand on the Earth you know that you are standing on a lump of rock that is not only spinning but moving around on the globe. It takes 23 hours, 56 minutes and 4.004 seconds for you to go around the world once. It takes 365 days for Europe to get 25 mm further from the USA. Yet the changes in the relative positions of the continents and the heights of mountains produced have an influence on climate. Don't hold your breath and wait for these processes to suddenly counter any current warming, but because it disrupts ocean and wind transfers of energy around the world, the movement of plates is another factor to add into the mix.

Don't Mention the Ozone Hole!

Global warming may be the current pan-global environmental worry but before it there was the ozone hole. British Antarctic Survey members discovered the hole in the ozone layer in 1985, and by 1987 the Montreal Protocol was enforced, banning harmful chlorofluorocarbon (CFC) gases. This rapid reaction to stop CFC production is a model for climate change policy makers. The levels of ozone in the stratosphere are expected to be back to pre-1980 levels by 2068. So why did this demonstration of united political opinion-achieving results not serve as a blueprint for global warming? Because, of course, there are no countries on Antarctica and the impacts of climate change are felt by many countries.

Is There a Link Between the Ozone Hole and Global Warming?

Up there with 'How many countries are there in the world?', this one is a dinner party minefield.

* If CO_2 is to blame, then a warmer lower atmosphere (troposphere) will create a colder upper atmosphere (stratosphere), which will increase the hole in the ozone layer.

* In a fascinating fight between cooling and warming, on balance increased ozone depletion leads to a cooler lower atmosphere (another one of those negative feedback loops).
* The ozone depletion greenhouse gases contribute 14 per cent to the estimated global warming increase from all greenhouse gases.

Having said all this, I would try to keep the two issues well apart as it only leads to some great student gaffes: '... the hole in the ozone layer lets more sunlight in and causes global warming...'

THE KYOTO PROTOCOL

Created in December 1997 and finally ratified in February 2005, the first and so far only global attempt to halt carbon emissions. Basing emissions on 1990 levels, most of the EU has met its targets without needing to indulge in carbon trading.

THE HUMAN

WORLD

WORLD POPULATION

Back in the twentieth century there was plenty of concern about the impact of population growth on the world and its resources. Could we support all the citizens of Earth? Were the great famines and environmental catastrophes an indicator of a world that was being stretched beyond its carrying capacity? Would the rich West have to give up some material wealth to help support the hungry nations? So, have things changed?

WORLD POPULATION GROWTH

UN global population estimate

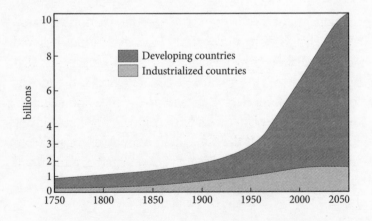

Like the number of bacteria in a Petri dish, the population of the world has increased close to exponentially since the Industrial Revolution. As the diagram shows, it is in the developing countries of the world that this growth is predicted to continue. This scenario should be a common enough memory from geography lessons. But is it still the case? Is exponential growth still going on? The UN seems to believe that the growth phase – be it exponential or arithmetical – of global populations will have ended by 2075. The world by then will, according to the UN, have a population estimated at 9.22 billion.

THOMAS MALTHUS

Thomas Malthus (1766–1834) was a Cambridge-educated economist turned clergyman whose views have influenced economic geography since the start of the nineteenth century. *An Essay on the Principles of Population* (1798 to 1826, six editions) remains the foundation for the pessimistic camp who believe 'We're all doomed!' In brief, Malthus believed that population growth would outstrip food supply and that would lead to 'Positive' and 'Negative' checks: the former consisted of hunger, disease and war – nice and cheery – while among the latter were birth control, abortion, celibacy and prostitution. If the population did not keep itself from reaching a resource-carrying capacity, then the result would be a catastrophic collapse in population.

ESTER BOSERUP

In 1965, the contrasting and more optimistic camp of economists and demographers who had long argued that humans would be capable of warding off Malthusian Crisis got their own standard bearer in the rather unlikely form of a Danish agronomist, Ester Boserup (1910–99). While working for the UN she produced the cumbersomely titled *Conditions of Agricultural Growth: The Economics of Agrarian Change under Population Pressure*. The reason she is held up as the foundation of the debate contra Malthusianism is that far from being concerned that humans would eat themselves out of food, her studies showed that as humans begin to see the empty shelves of the larder, they innovate to fill them again. In a nutshell: 'Necessity is the mother of invention'.

'Boserup's curve'

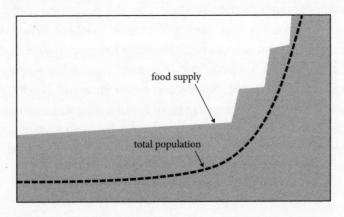

THE GREEN REVOLUTION

As the epitome of Boserupian responses, the Green Revolution is the name given to the massive increase in food yields in low-income countries (LICs) due to the application of technologies of farming from other countries. The phrase was first used in 1968, though the process had been ongoing since the end of the Second World War. The Mexican government, with financial assistance from the Ford and Rockefeller Foundations, sought to reduce food imports and to increase self-sufficiency.

The architect of the Mexican programme was Norman Borlaug, an American Nobel Peace Prize-winning agricultural expert. He also helped the Indian government and, with the International Rice Research Institute set up in the Philippines, helped to create IR8, a variety of rice that could produce more grains of rice per plant. IR8 revolutionized the farming of rice and, though there are many seed types utilized in different places around the world in growing yields, it remains synonymous with the Green Revolution. Critics say that it favours those farmers who can already afford the fertilizer and pesticides that the crops need to the detriment of the local-scale subsistence farmers.

GM CROPS

The very contentious issue of GM foods has to be balanced against their potential to increase food availability for the world. GM foods have in essence been artificially modified by genetic manipulation of organisms rather than just cross-pollination, as with IR8 (*see* above). The first GM crop was a tomato with the rather butch name of FlavrSavr, which, in 1994, was given approval to be sold by the US Food and Drug Administration. There is great concern still that there will be unforeseen side effects of letting GMOs into the world en masse.

THE NEO-MALTHUSIANS

Paul Ehrlich

His seminal work was the *The Population Bomb* (1968) and though his more recent writings have moved him away from the essential environmental-catastrophist school or doom-mongers, he is still the most often quoted Neo-Malthusian. Ehrlich put forward the view that the fundamental crisis for humans can be seen in the famines, civil wars and environmental catastrophes that have periodically dominated the news from the 1960s onwards. In the 1960s there was great concern about the rapid global population increase being seen across Asia – India and China being the most obvious examples. Ehrlich is a great example of the polymath geographer in that he trained in the field of zoology and applied his energies in this field as well as in the study of human-environment research.

Together with his wife, Anne Ehrlich, he continued to prick consciences on behalf of the loosely defined 'environmental movement' in his role as Bing Professor of Population Studies at Harvard University. As he and Anne wrote in *Population Bomb Revisited* (2009): 'The *Population Bomb* helped launch a worldwide debate that continues to today. It introduced millions of people to the fundamental issue of Earth's finite capacity to sustain human civilization.'

'The Club of Rome'

In a first attempt to blind with science, this international think tank of industrialists, diplomats and scientists, founded in 1968, commissioned Meadows, Meadows et al, to produce a report. They ran a computer simulation of what could happen to the world population if rates of resource use, pollution, population growth and other key indicators continued to change as they had been observed to do. The published report, *The Limits to Growth* (1972), gave new weight to the catastrophist Malthusians by projecting large-scale population decline as a result of the impact on the environment. The organization still exists and pinpoints global warming as a key battleground for humans to challenge themselves or risk the rather unpalatable consequences.

THE ECONOMISTS COUNTERING THE MALTHUSIANS

After Boserup raised her banner, it did not take long before a number of eminent economists and futurologists (can you imagine having that as your passport 'occupation'?) joined the debate.

Julian Simon

In many ways the arch-nemesis to Paul and Anne Ehrlich, Simon (1932–98) wrote his seminal piece *The Ultimate Resource* (1981) in response to his perception that one side of the population–resources debate had been presented too much as fact rather than a side in a broader discussion. The view put forward by Simon was that at times of scarcity the greatest financial gains could be made from innovation and substituting new resources so that entrepreneurs would fill the dangerous gap and either find more resources or produce alternatives.

THE BET

Famously, Julian Simon challenged Paul Ehrlich to
a simple bet that a basket of five metals would not
increase in price over a ten-year period. Though
Ehrlich did not want to take the bet for risk of giving
too much oxygen to a man he dismissed as a 'used-car
salesman', he named his five metals: copper, chromium,
nickel, tungsten and tin. None of these rose in price in
relative terms – indicating no scarcity: Simon won the
bet. Simon's scepticism won him fewer serious followers
and the environmental catastrophists continue to
occupy the greater share of the media.

Bjørn Lomborg

Having read a profile of Julian Simon in a magazine,
Bjørn Lomborg was inspired to write *The Skeptical
Environmentalist*. In it Lomborg appears to advocate
a more sensible recognition of the priorities facing
mankind. A wunderkind to many, a target of hate for
others, Lomborg often features in debates over climate
change as he advocates a more circumspect view of the
causes and mitigation of global warming. His essential

123

view is that we are better off promoting innovation as a way of accommodating all manner of ills than believing we can just stop all the nasties that the doom-mongers foretell.

POPULATION DISTRIBUTION

Where is the highest population density in the world? An age-old question that is worth a quick response. If you are looking for countries then here is the list:

Country	Population density (people/km²)
Macau (China)	18,534
Monaco	16,923
Singapore	7,023
Hong Kong (China)	6,349
Gibraltar (UK)	4,559
Vatican City	1,877
Malta	1,309
Bermuda (UK)	1,226
Bangladesh	1,127
Bahrain	1,099

If you really want to find extreme figures for population density you need to head into the densest parts of some big cities. Despite the idea of there being a 'City Limits' sign hung on the outskirts of many cities in the movies, for most, this does not exist. In addition, many cities are fluid in terms of their populations, with daily commuting mixed in with more permanent arrivals and departures. Hence any definitive list of cities by population is open to challenge.

City	Country	Population density (people/km²)
Mumbai	India	29,650
Kolkata	India	23,900
Karachi	Pakistan	18,900
Lagos	Nigeria	18,150
Shenzhen	China	17,150
Incheon/Seoul	South Korea	16,700
Taipei	Taiwan	15,200
Chennai	India	14,350
Bogotá	Colombia	13,500
Shanghai	China	13,400

IS TAIWAN A COUNTRY?

The answer is No – but why? In 1949, the island group known previously as Formosa became the retreat of a group of politicians, philosophers or just dissidents fleeing the new communist party of what we now know as The People's Republic of China (PRC). They declared Formosa and some smaller surrounding islands The

Republic of China (ROC). After the PRC took the ROC's seat at the UN the game was pretty much up. To avoid confusion, no one usually refers to the name ROC but uses the term Taiwan. The PRC considers Taiwan to be part of PRC territory, as does one of the two main political parties of Taiwan. Confusingly, the other party considers the state independent.

DEMOGRAPHIC TRANSITION

All countries go through a change in their birth rates and death rates, but there is only one Demographic Transition Model (DTM).

The Demographic Transition Model

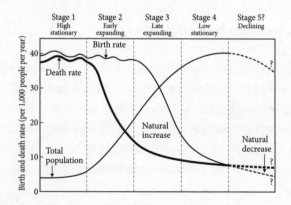

THE DTM

In 1929 Warren Thompson created this graph and so launched a thousand exam questions. Stage 1 is only really visible now in the more remote ethnic groups of Amazonia and South Eastern Asia. Historically this period preceded any agricultural or industrial revolution. Times were hard, many countries in Europe were feudal and death rates were high, meaning there was a lot of cavorting to secure heirs and hence the birth rate was high. These were the Middle Ages (fifth to sixteenth century): disease, crop failure and low hygiene standards.

After the Middle Ages and the Italian Renaissance (thirteenth to sixteenth century), came the Age of Enlightenment (Stage 2), when innovation and invention brought many changes. The most frequently quoted example of innovation in the UK was the seed drill, invented by Jethro Tull (more accurately, Tull improved upon a device that had been patented much earlier). This and other relatively straightforward innovations had a considerable impact. Gradually, agriculture improved, yields rose and basic sanitation increased access to clean water. Hence the death rate in the DTM starts to decline but there is a lag of cultural inertia before people notice the large number of children in the streets and change their habits or review their religious beliefs.

By Stage 3 there had been enough innovation and

change, and birth rates were now heading down as industrialization had started to influence society. Increasingly complex industrial cities across the globe had plumbing, antiseptic and the beginnings of female equality. Jump forward to the present day and countries that have led through demographic transition are now entering Stage 5.

When Thompson developed his model he could not have foreseen this change, and the challenge now is to predict what will happen next.

Stage 5 and the Decline of the Human Race

We started by looking at the classic Malthus versus Boserup debate. The catastrophe that Malthus predicted and the innovations that Boserup had discovered were both set against a belief in continued human population expansion. Well, Stage 5 shows us that we may have a bigger issue to contend with – underpopulation. If we have not got enough people to do work or enough people to buy things, then where will we stand? When birth rate drops below death rate a country will have a net loss of population.

The Total Fertility Rate (TFR)

If on average each woman in a country has 2.11 children in her lifetime and as long as migration is zero, then the population of the country will stay stable. So if we look at the number of countries in the world that have a TFR below 2.11 (or 2.30 in less developed countries) – the so-called replacement ratio – we will get an idea of what is going to happen to the world. You might want to forget about rising sea levels here and think about the next great catastrophe.

Global total fertility rate				
1965	2002	2009	2030	2050
Global* TFR				
5.0	2.8	2.55	2.1 (est.)	2.0 (est.)
* UN data at data.un.org				

It is now estimated that almost half of all countries in the world are below replacement ratio. Of course, if the great players in globalization keep moving their industries around the less developed countries, then those countries will industrialize rapidly and with that bring down birth rates. To put that in context, it means that world population will be declining by 2050, when sea levels will have risen by just 30 cm, it is predicted. So, fewer people just when we'll have less land. Of course, sea-level change is a very localized result of climate change but at a global level, the human system reduces the potential problem of sea-level change naturally. One could argue that this would be a Malthusian 'negative check'. When you start to bring together the myriad issues facing the Earth and its population you can begin to understand Professor Lomborg and his urging to consider the whole situation facing mankind rather than just climate change in isolation.

JAMES LOVELOCK

As the doyen of holistic approaches to global issues, James Lovelock (1919–) contributed much to the ideas of how we should frame the relationship between man and the environment. He first presented his hypothesis, or really his philosophy of Gaia, in 1972, when he was working for NASA. In essence, he and his supporters put forward the view that the world should be seen as a single, complex entity that was capable of homeostasis and, as such, could heal itself. The catastrophist side of this view is that humans and their actions have moved Gaia so far away from its ability to regulate and support itself that there is an inevitable point coming when it will collapse. The Gaia Theory, as its supporters now refer to it, has been heavily criticized by scientists from many fields, including Richard Dawkins and Stephen Jay Gould.

POPULATION PYRAMIDS

Strictly speaking, the analysis of population structure can be done by subdividing the data into any sub-group one chooses: gender, ethnicity, intelligence and so on. For most people the study of population structure means one thing, the population pyramid. To give it the correct name, these are age–sex relationship diagrams.

Classic Population Pyramids

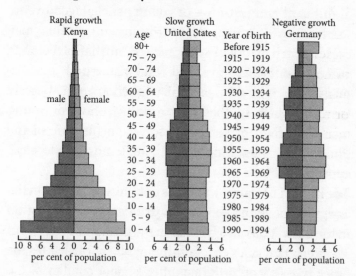

The population of a country is divided into five-year age cohorts and by gender. On the left is the wide-based expanding population pyramid of a country with a high birth rate (Stage 2 or 3 on the DTM, *see* p.126). In the middle is a country with a narrower profile but still a wide enough base to keep growth over the years (Stage 3 into 4). On the right is a pyramid of a country that needs migrant labour quickly as it is shrinking in numbers (Stage 5). The expanding pyramid is called that because that bulge at the base will pass on up the pyramid over time and thus expand the population. There are lots of things that show up on population pyramids:

Significant emigration – of young people (aged twenty to thirty-five) known as the 'brain drain'. This may be seen on a national scale such as in that of Poland's current population pyramid. Likewise, rural areas of many LICs will show a similar phenomenon. Usually there is a bias so that the working cohort of young men is under-represented; however, in the case of the Philippines, emigrating young female nurses affect the appearance of the pyramid.

The impact of wars – this leads both to a reduction in the age group of young people and to a temporary reduction in the base. It can also be seen in the post-war baby boom – a bulge that rises through the pyramid caused by the quick recovery of birth rates after a major conflict.

Significant social and cultural change – a short-lived increase in the birth rate caused by increased affluence, the post-Second World War 'boomers' reaching maturity in the 1960s, and a background of increased liberalism.

Immigration of economic migrants – typically aged twenty to thirty-five, possibly with children. On a population pyramid this bulge may remain in the same place, as very often the migrants return home and are replaced with others in the early part of the cohort.

DEPENDENCY RATIO

This equation helps to explain why there is a mounting problem for many affluent countries with ageing populations and a need to support them with public services.

$$\frac{\text{Dependency}}{\text{Ratio}} = \frac{\text{Number of dependent people}}{\text{Number of independent people}} \times 100$$

If you are over sixty-five years old you are deemed to be dependent. If you are reading this at your grandparents' house and are below sixteen years then you are most definitely a dependent too. All the remaining age groups are found below the line. As long as unemployment is not at a significant level then the denominator of the fraction is supporting the numerator above it.

Clearly, in countries with ever-ageing populations, which we used to call senile populations but are not allowed to now, the dependency ratio will be high. A ratio over 1 means there are more people relying on fewer people to pay for their services. From pensions to hospital places this simple equation holds the key to so many difficult political issues. Here are a few of the issues that feature in the political and media worlds of our time that relate directly back to the Dependency Ratio:

Migration – workers enter a country and go to the bottom line to help balance the ratio.

Private pensions and savings – this means that although the ratio does not alter, the actual lifestyle for dependents is better than expected as they are able to supplement the state pension support.

Pension age – of course, a sure-fire way to manipulate the situation is to move people from above to below the line. As most countries have laws to stop us moving the under-sixteens to the working age, we can only move the elderly group. Within the EU the modal retirement age is sixty-five, but there is considerable pressure to raise it to sixty-seven. In the USA the retirement age is already sixty-seven for those born after 1960. The situation in Russia, however, is stark, with a very high mortality rate* and a declining population; although the retirement ages are sixty for men and fifty-five for women, in reality people work way beyond them if possible.

---

* Russia has a shrinking population of 0.467% annually, ranking it 224 out of the 233 entities that feature in the CIA factbook in terms of population growth.

MIGRATION

Following on from the population pyramid the question of migration arises. It may be a 'thorny' issue but one can see the benefits of migration for the host country and, for some countries, the absolute need. Migration was and remains all about models and classifications.

Forced migration – persecution for religious, political or other beliefs, or for reasons of gender or sexuality, will make you a refugee. Under UN conventions any refugee can claim the right to asylum in a foreign country. The term asylum seeker is given to a displaced person who has not yet made a claim for asylum and is therefore not a refugee, or not yet. There are growing numbers of environmental refugees (victims of earthquakes, for example) but since 2007 the UN favours the use of the term 'Environmentally Induced Migrants'. In addition, those people forced to migrate from one place but who do not cross a national boundary are termed 'Internally Displaced Persons'.

Voluntary migration – all voluntary migration can vary over different timescales – for instance, commuting is an example of circulation and distances.

GRAVITY MODELS

Using Newton's theories of gravity, the expectation is that large settlements attract migrants more than small ones. In addition, the pull of the settlements decayed in the same way that the gravity of the Earth decayed, that is, exponentially. This idea of 'large bodies attract' is the analogy to the principles of such models, models that are of limited value in the twenty-first century.

LEE'S MODEL OF PLACE UTILITY

In 1967 Everett S. Lee introduced the idea that each voluntary migrant makes their own decision. Lee believed that before you migrate you make either a subconscious or a conscious assessment of both where you are and where you want to go. You create a list of the benefits and costs of each place and, provided the final score for the place you are going to is higher (the place utility), then you will migrate. Lee stated that this would only be the case if there were no obstacles – such as language, visa or financial issues – to making the move. In this case the likelihood would be that the migrant would go to an intermediate destination.

SETTLEMENT

The study of settlement has been a cornerstone of human geography ever since the subject itself got so large in its scope that geographers were forced into sub-subject areas. On that note, the last great true geographer who was able to provide a written synthesis of the whole range of the subject knowledge was Paul Vidal de la Blache, with his *Principes de géographie humaine*, published posthumously in 1923. After him the volume of research within this broad subject area grew too much for any one person. Hence settlement studies is one of the clear sub-groups of geography today. Of course, given the growth in urbanization and the impacts on rural settlements of modern trends in migration in low-income countries (LICs) and counter-urbanization in high-income countries (HICs), it is a vigorous field of geographical enquiry.

PAUL VIDAL DE LA BLACHE

Paul Vidal de la Blache (1845–1918) was responsible for the creation of one of the most individual of national geographical schools. The École Française de Géographie was remarkable because of its focus on regional rather than process geography. The contrast between the two schools is shown in this chapter, and in settlement geography as a whole, by the work of the German School and some of its key protagonists such as Weber and Christaller (*see* p.143). Through the founding in 1891 of *Annales de Géographie* with Lucien Gallois, Vidal de la Blache created an academic journal that focused on the holistic treatment of humans and their actions within the local environment.

SITE AND SITUATION

The study of the factors that affect the location and then growth or decline of settlements is a key place to start. Site refers to the local factors such as a bridging point or a natural spring, whereas situation relates settlements to each other. The easiest way to remember which is which, is that 'site' is a shorter word than 'situation' and thus refers to the smaller-scale factors of location.

SETTLEMENT MORPHOLOGY

Originally using maps, then moving on to aerial photography and now with the advent of GIS (geographic information system), and in particular Google Earth, settlement morphology – the shape of settlements – remains a key theme. Settlements can be broadly classified in several ways according to their shape. The broadest groupings are:

Linear – village follows a line such as a river terrace or a river itself, commonly found along route ways.

Nucleated – either a central focus such as a well, a green space or a crossroads, or the result of the 'three field system' (in which a triangle of fields with a settlement in the middle allowed citizens with strips of land in each field the most efficient access to each strip of land). Of course, in feudal England it did not escape the notice of the feudal lord that a nucleated population would be more readily available to raise arms or to pay tithes (in this case a tenth of crops harvested or money earned).

WHEN DOES A RURAL SETTLEMENT BECOME AN URBAN ONE?

In the old days it was easy to draw the distinction between urban and rural settlements: urban places had factories and police in riot gear; rural places had tractors and policemen on bikes. The car changed everything so that commuting became easier and even more people could buy second homes in the country. Now that rural people can shop in the edge of town and city retail parks, and urban people can bring their shopping out with them from their city at the weekend, the world has changed.

THE FIRST SETTLEMENTS

The Neolithic revolution at the end of the last Ice Age is the best place to start to find the settlements. In approximately 10,000 BP, plant and animal domestication started as the climate change brought joy to the world. In the lower Nile valley, the Indus valley and the Fertile Crescent (between the Tigris and Euphrates rivers in modern Iraq) we find the first evidence of settlements. In these idyllic farming areas of warm climates with river water and fertile alluvial soils it would appear that, as farming increased in yields, people were able to specialize in non-farming activities. It also meant an end to the constant wanderings in search of food of the old hunter-gatherer days in favour of the pizza-delivery of the modern world. Starting with the goat, dog and wheat the whole process of urbanization kicked off.

URBANIZATION

Statistically, the word urbanization simply refers to the population of a country living within urban areas. The United Nations includes the concept of rural to urban migration within its definition too. Whichever way one looks at it, it is just the expansion of urban dwellers over the rural dwellers, cities over villages. With it comes a shift in political and economic priorities, leading inevitably to the tyranny of the urban elite. In 2008 the world became officially urbanized, with more than half of the world's people living in urban areas. With 95 per cent of the global population living on only 10 per cent of the world's land, is it any wonder we feel squeezed?

WHEN IS A VILLAGE NOT A HAMLET?

As every country has its own definition of what constitutes a settlement of each size, it is almost impossible to make country comparisons using local classification names. For example, a town in Denmark may have only 250 people in it, whereas in Japan it would require 30,000 to get the title.

The normal Anglo-Saxon hierarchy of settlements is:

Isolated dwellings – one or two buildings with no clear settlement focus.

Hamlet – population below 100 and few if any services.

Village – in various sizes; from a one-shop village of 100 to large dormitory villages exceeding 1,000.

Town – again, of various sizes but with many services and up to 100,000 people.

City – again, a large variety of population sizes but with enhanced services.

Conurbation – when two or more cities grow and morph into one whole urban entity, usually with a population of several million.

Megalopolis – French geographer Jean Gottman first used this term to describe the Eastern Seaboard of the USA (having borrowed the phrase from *The Culture of Cities* [1938] by Lewis Mumford). It is usually restricted to the classic megalopolises of:

> San-Francisco – on the western coast of the USA
> 'BosWash' – the Eastern Seaboard of the USA
> Tokyo-Osaka-Kobe – where more than half of the Japanese population of 127 million live.

But with rapid industrial and urban growth, more areas are rising to claim the title. Although Mumford originally thought this label had negative connotations for the settlement and people living therein, more often now it is a badge sought with national pride.

CHRISTALLER AND HIS EFFICIENT HEXAGONS

As described earlier, whilst Vidal de la Blache created a coherent French School of Geography that focused on the interactions of the environment and people, the Germanic school was very different. By limiting the significance of natural features in their models, the German school was able to analyse the human factors responsible for various phenomena: Alfred Weber (1868–1958) and industry, Johann Heinrich von Thünen (1783–1850) and agriculture, and then there was Christaller. Walter Christaller (1893–1969) remains the godfather of settlement distribution study. Having studied the layout and hierarchy of settlements in rural southern Germany, Christaller developed Central Place Theory. His work attempted to explain why some settlements grew in importance whilst others in the same region did not. As the diagram illustrates, he explained his theory with the help of the great hexagon patterns you may recall having to learn to reproduce.

Christaller's k = 3 network

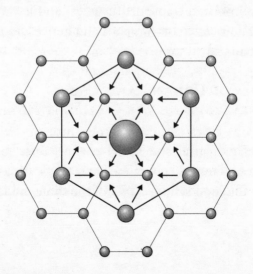

The essence of the idea is that a central settlement will prove to be the most efficient place to take advantage of all the consumers in an area. In the diagram, above each central place claims a third of the consumers for the settlements around it in the next lower tier of the hierarchy.

$$1/3 + 1/3 + 1/3 + 1/3 + 1/3 + 1/3 = 2$$

If you then include the consumers from the central place itself, you get 3. Christaller called this his *klasse* = 3 network and used it to explain why the settlements in rural areas will have grown into their local hierarchy. This was the most efficient arrangement for the settlements, so that both consumers and salesmen (mainly farmers in his model) could maximize choice and market size. Christaller put forward two other pretty patterns: k = 4, which minimized transport distance, and k = 7, which ensured no overlap of influence and hence was ideal for efficient administration.

MODELS OF URBAN LAND USE

As soon as cities started to grow through the various stages of industrial and social change, their form had interested planners, governors and the middle classes looking for good schools. Two of the three models most studied come from the Chicago Group and represent the application of social ecology to cities. Dating from the first part of the twentieth century, they are riddled with limitations and are almost useless when applied to new cities in countries like China and Brazil. But as a place to start they remain important.

E. W. BURGESS'S CONCENTRIC ZONES

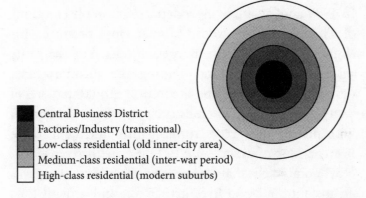

■ Central Business District
■ Factories/Industry (transitional)
■ Low-class residential (old inner-city area)
■ Medium-class residential (inter-war period)
□ High-class residential (modern suburbs)

In 1925 sociologist E. W. Burgess produced his simplified analysis of how different groups in a city arrange themselves across the land. Although based only on US cities (almost wholly on observations of just four cities, and disproportionately on Chicago), the model will be familiar to most readers. Burgess relied heavily on bid-rent theory (*see* p.147) to build his model. Using ideas borrowed from plant ecology in this group of social ecologists, Burgess suggested that through financial ability to pay, different income groups would colonize areas available to them. Though not rocket science to those who know 'the places to live' in London or New York, this application of ideas from one field to another was pure Burgess.

HOMER HOYT

In 1939 and, again, heavily dependent on observations of Chicago, the economist Homer Hoyt produced his slight reworking of the Burgess model. For the same reasons of bidding and gaining the location most desirable, he mapped the theoretical growth pattern of a city. For example, in cities such as Paris and London, with picturesque rivers running through them, there is a corridor of extra value to homes with river views. Hoyt pointed out that for those who had a car there was an instant ability to live further out and growth thus followed route ways. Hoyt was most closely linked to the correlation between ethnic group and location and explaining ethnic ghettoes. One sure thing is that as flooding becomes more widespread there will be plenty of people selling their waterfront homes and influencing more urban change.

Hoyt Sector Model

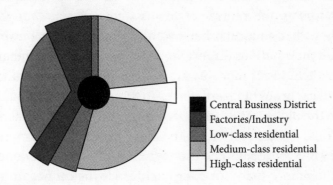

Central Business District
Factories/Industry
Low-class residential
Medium-class residential
High-class residential

BID-RENT THEORY

This is the rocket science behind Burgess's model.

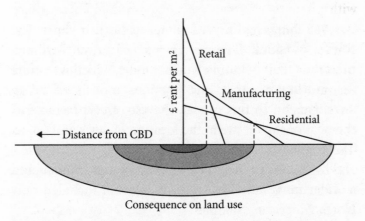

Consequence on land use

The diagram shows the ability for different functions

to bid to rent locations relative to the distance from the centre of the city. Before the age of out-of-town shopping, the centres of towns were where it was at. Shops had to locate there as the omnibus and the train had their stations disgorging shoppers there. Remember that the classic models were constructed well before the evil car brought its influence. To the edge of town would be the old factories and industrial sites, and the workers would live around them. The more you earned the more readily you could flee the stench of the inner city and enjoy more space for your buck. Since the car began to exert its hold on people and road builders, things have changed a great deal.

HARRIS AND ULLMAN SEE THE FUTURE . . .

By 1945 things had moved on apace. In their paper 'The Nature of Cities', Chauncy Harris and Edward Ullman presented their Multiple Nuclei model, which was more forward-looking than the previous models of urban development. In this model, the use of cars has altered the relationship between the location of work and home. The result is the growth of outlying business districts and high-class residential areas. Harris and Ullman used a wider range of cities to study but still included only North American examples.

The Harris and Ullman Model

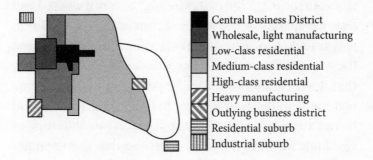

Central Business District
Wholesale, light manufacturing
Low-class residential
Medium-class residential
High-class residential
Heavy manufacturing
Outlying business district
Residential suburb
Industrial suburb

NEW URBANISM

A movement only thirty years old, New Urbanism seeks to review cities and the way they are run so that they are not driven by the car and bid-rent theory, but by ideas of sustainability and quality of life for the inhabitants. By bringing together architects, planners, developers, politicians and academics, the movement is growing in influence and acts as an umbrella to include people outside the immediate 'membership', such as Jaime Lerner.

JAIME LERNER AND CURITIBA

Though the former governor of the state of Paraná, Brazil, Lerner rose to prominence with his activities in Curitiba, the state capital. An urban planner and architect, he introduced a host of quite straightforward schemes to the city with the intention of improving the way it worked and thus the quality of life for its inhabitants. His first focus was on transport; he reorganized and then prioritized the bus service to allow more Curitibans to use it rather than their cars. In addition, he pedestrianized shopping areas and introduced a domestic recycling scheme that rewarded people, often in the poorest areas, with food in exchange for the volume of recycling that they brought to the refuse trucks. He has since gone on to promote many of the simple but effective ways that cities can be transformed for the better.

WORLD CITIES

The idea of the World Cities dates to 1915, when Patrick Geddes (1854–1932) used the phrase in his *Cities in Evolution* in conjunction with a small list of what he considered to be the highest level of interaction between cities at a global scale. Typically, these cities are centres of cultural, political and economic importance beyond their national boundaries. In 1998 the GaWC (Globalization and World Cities Study Group Network) published the first definitive study of the places.

City	GaWC rank (previously alpha++ = 1; alpha + = 2; alpha = 3)	Global City Index (from 2008 *Foreign Policy*, American journal based on 25 statistical measures)	Global Power City Index (2009 publication from the Mori Foundation of Japan)	Total Population (million)*
New York	1	1	1	3 (22.3)
London	1	2	2	29 (8.6)
Paris	2	3	3	22 (10.5)
Tokyo	2	4	4	1 (34.7)†
Hong Kong	2	5	10	42 (7.0)
Singapore	2	7	5	65 (4.5)
Shanghai	2	20	21	14 (14.7)
Beijing	2	12	26	18 (12.8)
Sydney	2	16	14	81 (3.7)
Moscow	3	19	32	16 (13.7)
Seoul	3	9	12	7 (19.7)
Madrid	3	14	11	52 (5.4)
Milan	3	39	29	68 (4.3)
Brussels	3	13	18	200 (1.7)
Toronto	3	10	15	50 (5.4)
Mumbai	3	49	34	1 (13.9)
Buenos Aires	3	33	n/a	59 (3.0)
Kuala Lumpur	3	40	27	66 (4.1)

*Includes the urban fields of each settlement to aid country comparisons.
† Includes the Tokyo-Yokohama population.

So, is there a definitive list of World Cities? Well, obviously not. How, after all, does one pit the economic powerhouses of London or New York against the sheer joy of living in Sydney?

INDUSTRY AND ENERGY

SECTORS OF INDUSTRY

A revolution started in the last few years with the appearance of a 'quinary' industrial sector. I'm sure that in the days before the Internet and the mobile phone we only had four sectors and before that, if you can stretch back that far, three.

The Clark Model

In the Clark–Fisher model, the three broad sectors of industry – primary, secondary and tertiary – are shown to change as a country develops. The advent of more modes of working is the cause of the accretions to this model.

Primary sector – extracting of raw materials, including farming and fishing.
Secondary sector – manufacturing of goods from raw materials, from making widgets for beer cans to produce good heads (a true demonstration of the height of human

research) to gluing bits of computers together in Manila. **Tertiary sector** – to you and me, the service sector. As long as a tangible product was not being made it slipped in here. From prostitutes to football mascots, a wide range of moral and less moral pursuits could be found in this category.

The quaternary sector was added – which just meant Research and Development to everybody. The quinary sector is demarcated as being the service industries of health, some education and research. Which leaves intellectual activities – government, culture, libraries, scientific research, education and ICT. This leaves all the other non-primary and non-secondary industries in the quaternary or tertiary sectors.

THEORIES OF INDUSTRIAL LOCATION

Up there with von Thünen and Christaller, Alfred Weber was another of those great German economist–geographers who applied numbers to modelling what he saw around him. In the days before microchips and mobiles, Weber invented the Material Index.

WEBER'S MATERIAL INDEX

In a simplified vision of the world, Weber considered

a factory with one market and two raw materials. If the owners of the factory were to make the most profit they would need to find the Least-Cost Manufacturing Location. If the industry had two heavy raw materials, Weber predicted that the LCML would be closer to their sources to reduce overall transport costs (this helps to explain the early location of iron and steel manufacturing in the UK near the coalfields of South Wales and the Midlands, and the coal and iron ore deposits of the Ruhr valley in Germany). As these initial resource areas were exhausted, raw materials from abroad became cheaper to import. This was helped by the increase in the size of Very Large Crude Carriers (VLCCs) so that now coastal locations had the competitive advantage. Although Weber was dealing with a completely different world of industrial production, his model can be adjusted to explain the modern changes quite readily. Such an industry would have a Material Index of greater than 1 – in that weight was lost during production and hence the industry could be termed material-orientated. The opposite type of industry would be a market-orientated industry, a good example being brewing.

INDUSTRY IN THE DEVELOPING WORLD

Transnational companies, reduction in transport costs, the stabilization of national politics, trading blocs and the increased size of ocean-going transport have all helped to facilitate the growth of all kinds of industry in the low-income countries. Traditional theory that said that primary and secondary industry would first move to the LCML has been eclipsed as the Internet has allowed all manner of tertiary jobs – from selling flight tickets and mortgages to customer services operations for credit card companies – to be outsourced. India has led the field in capturing outsourced jobs in the tertiary sector. In addition, India has seen a growth of its own transnational companies, led by Tata, which through a series of acquisitions now owns famous industrial brands across the world.

THE FORMAL AND THE INFORMAL SECTORS

Whereas the formal sector of employment includes strong regulation and protective rights for workers, the informal sector does not. Though because of its nature the scale of the informal sector is difficult to quantify, it is estimated that nearly 75 per cent of workers in sub-Saharan Africa work in the sector, and in urban unplanned settlements (shanty towns) across the LICs that figure is thought to rise to nearly 100 per cent. The dilemma for national governments is whether to risk the social upheaval and potential drop in many people's standard of living by trying to regulate this work so as to gain the taxation that may be used for long-term development, or to accept that the revenue not made on formalizing the employment is the cost worth bearing for the people involved. Certainly, many countries stagnate in terms of Rostovian development (*see* Rostow's Model, p.177) by not formalizing their informal sector.

ENERGY RESOURCES

The First Law of Thermodynamics states that energy can neither be created nor destroyed. There are two basic energy resources – stocks and flows. Normally, these are labelled as Non-renewables and Renewables. The fossil fuels made up of the crushed, squeezed, heated-up remains of plants and animals from millions of years ago are a biological store of energy. On combustion that energy is released and that released energy is captured and utilized. Be it heating steam to generate electricity or converting it into the kinetic energy of piston rods in a car engine, the principle is the same. The problem, of course, is that we are using up the world's stock of this energy source. By contrast, there is energy in the Earth's crust (geothermal), in the oceans (tidal, wave, ocean current) and in the atmosphere (wind and solar) that is not a stock resource and therefore will not be depleted. The essence of renewable energy is to take advantage of the flow of energy to release dependence on the ever-depleting stock sources. In the light of this and local issues of fuel security there are various targets around the world for reducing non-renewable use: for example, the EU target is to have 20 per cent of energy to come from renewables by 2020.

TOURISM

LEISURE VS RECREATION VS TOURISM

The largest industry in the world. If you are interested, the difference between leisure and recreation is *activity*: while leisure involves anything that is not work, recreation implies that you are doing something. The World Tourism Organization defines tourists as people who 'travel to and stay in places outside their usual environment for more than twenty-four hours and not more than one consecutive year for leisure, business and other purposes not related to the exercise of an activity remunerated from within the place visited'.

THE HOTTEST TOURIST CITIES

International tourism is a notoriously fickle entity but the figures in the table overleaf show a list of cities that are most frequently to be found in the top ten.

City	Number of visitors (figures for 2007/2008)
Paris	15,600,000
London	14,800,000
Bangkok	10,840,000
Singapore	10,100,000
New York City	9,500,000
Hong Kong	7,940,000
Istanbul	7,050,000
Dubai	6,900,000
Shanghai	6,660,000
Rome	6,120,000

THE GROWTH OF TOURISM

There has been a dramatic rise in tourism around the world since the Second World War. Once people at the top of any new civilized society have others beneath them to provide food and wealth, they have leisure time. Thus the Incas, the Maya and the Pharaohs certainly had time for a quick game of Trivial Pursuit while others looked after the place. The real claimants to the throne, though, were the Romans and Greeks. The Romans famously had their second holiday homes in the cooler hills around the city, with bags of leisure time to pursue what can only be described as unusual forms of relaxation.

Global tourism in terms of travellers

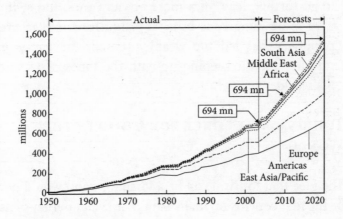

Tourism on all scales has rocketed since the end the 1950s and for quite obvious reasons:

* Increase in people with paid holidays.
* Increase in average disposable income (make that income for leisure).
* Improved transport technologies – reducing the friction of distance.
* More roads with higher speeds.
* Increased car ownership.
* Expanded air travel including post-war trained pilots and plenty of planes for converting from military to civilian use.
* Developments in the travel industry.

* The media has undoubtedly fuelled the desire of many to go further, leap off a more obscure rock and swim with more dangerous fish. This hunt for the pleasure periphery has led the wealthy retired, the gap-year traveller and the keeping-up-with-the-Joneses to scour the world.

TOURISM AS A FORCE FOR GOOD IN THE WORLD?

The idea that tourism can be a force for good would not be the first conclusion many would draw but perhaps it is useful to occasionally take an optimistic view of things.

The economic issues

On the positive side, bringing in foreign cash must be a good thing as it will kick-start the local 'multiplier effect' and contribute to GDP. Even if some of the tertiary activity is not of the most wholesome variety. It's worth adding in here that it may give the local rural people an increased market for their food, a more hidden benefit as it stops people heading to the cities. Anything to reduce the rate of urbanization in LICs must be beneficial?

The economic downside may include inflated prices for everything from food to land, which the local people will have to pay. Secondly, there has to be investment in the infrastructure needed to support tourism. But then again if some of that infrastructure increases available power supply or access to clean water, there will be other winners. By far the biggest economic negative is a comparative one – leakage. If you stay in an international hotel your profits may be leaked back and therefore taxed in the host country. The managers in your hotel may be well-trained company men and women, paid in dollars into their American bank accounts. Finally, there's product leakage. For example, the world's most global beer (the one sold in the most countries of the world) is Heineken, and given that the majority of tourists are from Europe, that's what a lot of them will want to drink, so the beer has to be imported or brewed under licence.

Aside from these points, there lurks the spectre of dependency. Countries using tourism as a means of development leave themselves vulnerable. Be it because of a global recession, a reluctance not to fly due to terrorist threat or simply the discovery of new places to holiday, through no fault of their own these countries may find that the tourists are not arriving, the disco isn't pumpin' and the streets are not teeming with people.

The environmental issues

To some, redeveloping a quaint Greek fishing harbour or chopping up a stretch of mangrove to make way for an all-inclusive enclave resort would seem mad. Unfortunately, there's gold in them there tourists! The debate between the environmental impact and the economic value of tourism is the most repeated in the subject area. Suffice it to say that from the top-down Tunisian government tourist industry to the bottom-up CAMPFIRE (Communal Areas Management Programme for Indigenous Resources) project of Zimbabwe, the balance is not always in favour of the environment. But here is where things start to look good, as the tourist industry has found its latest niche – ecotourism.

The cultural issues
There will come a time in a far-off age where we will all be true global citizens. Nourished by McDonald's, Coca-Cola and Domino Pizza we will drive our Fords up to our BAA airport, get on our Virgin Airways flight to stay at our Holiday Inn. While doing this we will all, of course, be listening to Mariah Carey on our iPods. And lo, we will have truly been globalized.

MODELLING TOURISM

The only real model or theory that you met if you studied tourism may have been Butler's Model.

Butler's Model

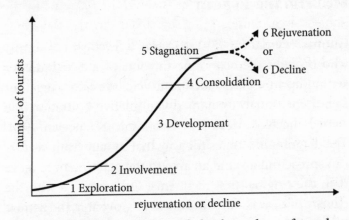

If you graduated into any of the branches of product marketing you will know this as the Product Lifecycle. Best illustrated by being applied to holiday resorts, it literally is nowhere near rocket science. Stage 1: a resort is 'discovered' and as more people arrive, the local community starts to take advantage and gets involved (stage 2). Over time, the destination starts to grow in popularity as the facilities at the location are developed. This is the 'golden age' for many resorts, as visitor numbers rise and takings increase. But it is in this period of consolidation that many resorts miss out on preparing for the next stage and the results of more competition and changing tastes. While some resorts may change and rejuvenate well, others stagnate or decline.

CUMULATIVE CAUSATION OR 'THE MULTIPLIER EFFECT'

Gunnar Myrdal (1898–1987) was a Swedish economist who brought the economic impacts of a new industry setting up in a place, such as an enclave hotel, together in a theory correctly named Cumulative Causation. He shared the Nobel Prize in Economic Science in 1974. The diagram illustrates the way that money from visitors can spread out around an area to the betterment of more than the original first recipients. Of course, as with the closure of a car factory in Detroit or Coventry, the demise of a location as a tourist attraction can happen rapidly and through no fault of the people there. It was for this reason that appeals went out quickly for tourists not to shun areas hit by the tsunami of 2004. The revenue from tourists can be fickle but as a form of aid it is vital. This is also the reason why, despite massive destruction in Port-au-Prince following the earthquake of 2010, cruise ships were being begged to carry on visiting the northern coast of Haiti.

Unfortunately, this little nugget seems to be available mostly as its nasty cousin – the Negative Multiplier Effect.

The Multiplier Effect

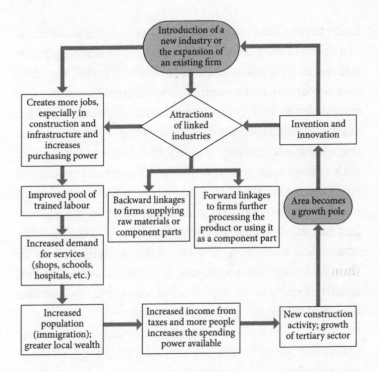

DEVELOPMENT

In the days of the Cold War (1945–91), the global divisions of nations were perfectly reflected in the labels given to nations of different wealth. The 'Third World' label was a convenient way of referring to the countries that were not Western Capitalist pact members or Communist pact countries. Thus the First World consisted of the developed, democratic countries, the Second World were those with communist governments and the rest were lumped together as the Third World. The use of the term 'Third World' is an unfortunate and lazy practice in the twenty-first century but it persists in the media and in the majority of people's vocabulary. The inherent hierarchy within the division was made redundant once globalization helped countries to develop rapidly and, of course, once many communist countries began to change.

NORTH–SOUTH DIVIDE

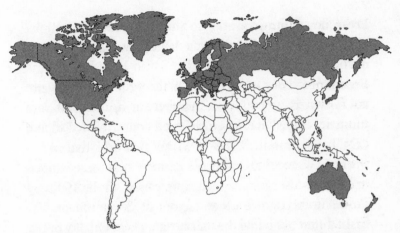

Willy Brandt, Chancellor of West Germany 1969–74, first proposed the idea of what became known as the Brandt line in the 1970s. The Brandt line divided the map of the world in two. The rich countries in the north, the poorer countries in the south. The line is still used today even though developments in many countries have altered the exact position while retaining the overall concept.

The labels of development and relative under-development have changed over the years as liberalism, political correctness and ultimately common sense have had their influence. I will tread the minefield of this rebranding with a snapshot of how we got to high-income countries, low-income countries and middle-income countries. Yes, the modern world is now full of HICs, LICs and MICs.

From 'developing country' to South country – the stigma attached to 'developing' as a label stimulated Brandt's change of language.

From South country to LDC – in recognition that poverty is a relative issue, countries changed during the 1980s into more developed and less developed countries (MDC and LDC together in the LDW or MDW, or world, that is).

From LDC to LEDC – in recognition that there is more to life for most humans than simple cash, in the 1990s we moved to recognize other factors of development. The main difference being the measurements of Quality of Life being raised in recognition rather than the more usual indicators of Standard of Living. As a result, all textbooks had to be reprinted and all teachers re-programmed to refer to LEDCs (Less Economically Developed Countries), MEDCs and NICs. The Newly Industrialized Countries account for those that, since 1950, have been steaming ahead with economic development, such as South Korea and Malaysia.

From LEDC to LIC – only more recently has the new naming appeared on the ugly horizon of Exam Board specifications to teachers. It would appear that the world continues to change and pen pushers at the UN or at National Statistical centres have come up with a new reason to keep their jobs. This is a UN World Bank classification based on Gross National Income. GNI is like good old GDP, but includes the interest and dividend payments paid into the country (less the ones paid out, of course). As the world has got more globalized in terms of business, this statistic looks set for a good run in general usage.

Current UN groupings

Classification	Countries
High-Income Country (HIC, $11,906 or more) (NB: the Isle of Man is counted as a nation in the statistics because of its tax foibles)	Andorra, France, Netherlands Antilles, Antigua and Barbuda, French Polynesia, New Caledonia, Aruba, Germany, New Zealand, Australia, Greece, Northern Mariana Islands, Austria, Greenland, Norway, Bahamas, Guam, Oman, Bahrain, Hong Kong (China), Portugal, Barbados, Hungary, Puerto Rico, Belgium, Iceland, Qatar, Bermuda, Ireland, San Marino, Brunei Darussalam, Isle of Man, Saudi Arabia, Canada, Israel, Singapore, Cayman Islands, Italy, Slovak Republic, Channel Islands, Japan, Slovenia, Croatia, Republic of Korea, Spain, Cyprus, Kuwait, Sweden, Czech Republic, Liechtenstein, Switzerland, Denmark, Luxembourg, Trinidad and Tobago, Estonia, Macao, China, United Arab Emirates, Equatorial Guinea, Malta, United Kingdom, Faeroe Islands, Monaco, United States, Finland, Netherlands, Virgin Islands (US)
Low-Income Countries (LIC, $975 or less)	Afghanistan, Guinea-Bissau, Rwanda, Bangladesh, Haiti, Senegal, Benin, Kenya, Sierra Leone, Burkina Faso, Democratic People's Republic of Korea, Somalia, Burundi, Kyrgyz Republic, Tajikistan, Cambodia, Lao People's Democratic Republic, Tanzania, Central African Republic, Liberia, Togo, Chad, Madagascar, Uganda, Comoros, Malawi, Uzbekistan, Democratic Republic of the Congo, Mali, Vietnam, Eritrea, Mauritania, Republic of Yemen, Ethiopia, Mozambique, Zambia, Gambia, Myanmar, Zimbabwe, Ghana, Nepal, Guinea, Niger

Middle-Income countries (MICs, from $976 to $11,906; these are split into two subdivisions: Lower MICS and Higher MICS, but here are united as one group)	Albania, Honduras, Paraguay, Angola, India, Philippines, Armenia, Indonesia, Samoa, Azerbaijan, Islamic Republic of Iran, São Tomé and Principe, Belize, Iraq, Solomon Islands, Bhutan, Jordan, Sri Lanka, Bolivia, Kiribati, Sudan, Cameroon, Kosovo, Swaziland, Cape Verde, Lesotho, Syrian Arab Republic, China, Maldives, Thailand, Republic of the Congo, Marshall Islands, Timor-Leste, Côte d'Ivoire, Federated States of Micronesia, Tonga, Djibouti, Moldova, Tunisia, Ecuador, Mongolia, Turkmenistan, Arab Republic of Egypt, Morocco, Ukraine, El Salvador, Nicaragua, Vanuatu, Georgia, Nigeria, West Bank and Gaza, Guatemala, Pakistan, Guyana, Papua New Guinea, Algeria, Grenada, Peru, American Samoa, Jamaica, Poland, Argentina, Kazakhstan, Romania, Belarus, Latvia, Russian Federation, Bosnia and Herzegovina, Lebanon, Serbia, Botswana, Libya, Seychelles, Brazil, Lithuania, South Africa, Bulgaria, Republic of Macedonia, St Kitts and Nevis, Chile, Malaysia, St Lucia, Colombia, Mauritius, St Vincent and the Grenadines, Costa Rica, Mayotte, Suriname, Cuba, Mexico, Turkey, Dominica, Montenegro, Uruguay, Dominican Republic, Namibia, Bolivarian Republic of Venezuela, Fiji, Palau, Gabon, Panama

Notice that to avoid any nation being upset with the way I have presented this table I have eschewed the practice of placing LICs third. With the MICs being subdivided already into Higher MICs and Lower MICs, how long before we get a revision of this classification?

The BRIC Nations

In the sphere of diplomacy and geopolitics there is one acronym that has emerged ever since our schooldays to put fear into the 'old world order'. Brazil, Russia, India and China make up the BRIC countries. Originally identified as a coherent group by Goldman Sachs in 2001, the name has been readily assimilated into the media as the belief is that this group of nations will have a combined wealth greater than that of the G6 by 2050.

WHAT IS DEVELOPMENT?

Measuring Development

It may be easy to compare two siblings as they grow up: 'He's smarter than his brother'; it may be easy to compare two or more sprinters too: 'They are all slower than Usain.' But how do you compare different countries with different cultures and different values, problems and resources? The answer is statistics.

Standard of Living

Just how useful is it to know which country is ranked eighth in terms of percentage of national population with mobile telephones? (It's the UK with a staggering 123 per cent, which really shows up the uselessness of such statistics!) If you are someone who finds this useful, then I'm sure you are a strong believer in measuring every apple and every orange to compare them. GDP per capita is still the most commonly used indicator of standard of living even though more recently it has been replaced by the GNI per capita. All such measures reflect the economic condition of the country and hide inequalities within the society. For that reason these indicators are usually now balanced against those of Quality of Life.

Quality of Life

The thousands of magazine adverts for overseas charities tell us that images speak more about the conditions for people in a country than the spare statistics do. Quality of Life indicators attempt to get close to the ability of images in giving a glimpse into the lives of people in a country. The table demonstrates the variety of performance for twenty countries chosen to represent the range of values for each criterion. No system of randomization, systematization or stratification has been used in the selection process.

Country	Standard of Living		Quality of Life		UN composite indicator
	GNI per capita	GDP per capita	Infant mortality*	Adult literacy†	HDI‡
USA	42,000 (6)	44,000 (8)	6.3 (180)	99.0 (19)	0.956 (13)
UK	37,000 (10)	39,000 (13)	4.8 (193)	99.0 (19)	0.947 21)
France	35,000 (17)	37,000 (18)	4.3 (170)	99.0 (19)	0.961 (8)
Australia	35,000 (18)	37,000 (15)	3.3 (217)	99.0 (19)	0.970 (2)
India	731 (146)	817 (160)	30.2 (103)	61 (147)	0.612 (134)
China	1,700 (120)	2,000 (131)	20.2 (105)	93.3 (80)	0.772 (92)
Bangladesh	445 (167)	429 (183)	59.0 (39)	47.5 (164)	0.543 (146)
Malawi	157 (190)	170 (205)	89.0 (14)	64.0 (146)	0.493 (160)
Burkina Faso	390 (171)	457 (182)	84.5 (15)	23.6 (177)	0.389 (177)

*Defined as: the number of infants born alive who die in their first year. *CIAFACTBOOK* 2009.

†UNESCO gives one standard: 'the percentage of population aged 15 years and over who can both read and write with understanding a short, simple statement on his/her everyday life'.

‡The Human Development Index is the current index of choice when comparing countries. This composite index gives a range of values down from 1.000. It includes elements of: life expectancy, adult literacy rate and Gross Domestic Product. Launched in 1990. UNDP 2009.

The positions indicated in the brackets are for all countries and therefore exceed the normal 193 full UN member countries.

THE NUMBER OF COUNTRIES IN THE WORLD

As can be seen from various lists in this book, there is no one definitive answer to the question of how many countries there are in the world, be it because some states are actually principalities or because there is genuine dispute over a country's governance. Here is a summary of the situation.

Agency or Group	Number
UN*	192 full members + the Vatican City = 193
CIAFACTBOOK †	266
US Department of State	194
'Sovereign States'	203

*This list is longer as it contains those countries which have claimed sovereignty but which the wider world is a little too nervous about recognizing.

†*CIAFACTBOOK* has the largest figure as it chooses to sub-divide the world into 'world entities' such as, for example, the Isle of Man.

STAGES OF ECONOMIC DEVELOPMENT

The idea that all countries would go through the same processes and stages of development as had been experienced by the UK, Europe and the USA is for many implicit. The good news is that despite their lowly position in the table of indicators, the Indians use their own rocket and satellite technology to track shoals of fish to help subsistence fishermen. In other words, you can jump stages by copying the wheel, not reinventing it. It was the Japanese after the demolition of their industry at the end of the Second World War who smartly spotted that they could do contemporary development their own way.

ROSTOW'S MODEL OF ECONOMIC DEVELOPMENT

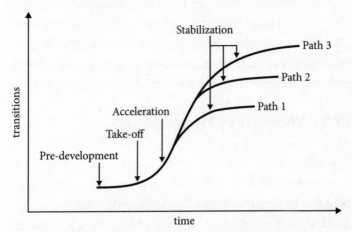

Walt Rostow bestowed this neat model on the world in 1960. Based on the experiences of European countries and the USA, the model shows how these countries developed and is often used to explain the route that other countries will take. So why is it still called the 'aeroplane model'? Quite simply, from the first textbooks until the present day each stage has been illustrated with a different type of aeroplane:

* Pre-development – Kitty Hawk flyer
* Take-off – First World War biplane
* Acceleration – Boeing 747
* Stabilization – Concorde

Crucially, Rostow claimed that a country needed to re-invest 5 to 15 per cent GNP annually to move through the crucial Acceleration stage. Even though there are various criticisms that can be made, this idea of the need for re-investment helps to explain the reasons why many LICs do not readily develop. Debt repayments, disaster, emergencies and large population growth all eat into any potential re-investment by either public or government.

ENVIRONMENTAL DETERMINISM

Traceable back to Strabo (63 BC–AD 24), this is an explanation that is still held by some in analysing the origins of and reasons for underdevelopment. Put most simply (and most disagreeably), the environment determines the people and their culture. The idea (let us not grace it with the term 'theory') was largely discredited in the 1920s but some elements of it last as an implicit explanation for many issues.

AID

Like chocolates in a box, Aid comes in many flavours, the luscious bounty of the fudge-filled centres being the work of NGOs, the lurking nastiness of the orange crème being the Tied Aid. Whereas NGOs can be reliable, cheap and able to work in dispersed rural areas without fomenting revolution, Tied Aid does exactly what it says on the tin. Tying in future spending to the donor country always has an unpleasant whiff to it, but often there is little choice and mutual advantage.

NGOs – Non-governmental organizations or charities are often the first agency that people think of when they think of aid. Estimates of the number of international NGOs in the world are about 40,000, while national figures vary from 170,000 for the UK to nearly 2 million active in India. Politically unaligned, their neutrality is important for them to get access to help people.

Tied Aid – Essentially the donor government or governments put a condition on the gift of aid that any services or goods must be bought from them. Pleasingly, this type of aid is being phased out, with the UK abolishing it in 2001. In the same year, the OECD started a process which ended on New Year's Day 2002 eliminating tied aid.

Bilateral Aid – Given from one country to another. This aid finds itself typically within an ex-colonial relationship. By way of example, the UK gives bilateral aid to India of £825 million over three years. Clearly, there is a sense of historical reparations underpinning much of this giving.

Multilateral Aid – USAID is as familiar to twenty-four-hour news watchers of a certain age as MASH. You may not, however, be so familiar with EuropeAid. This is the title of the European Commission's arm for organizing EU external aid programmes. The EU is the largest bilateral or multilateral donor if you exclude the United Nations, channelling 60 per cent of world aid annually.

THE IMPACT OF AIDS IN DEVELOPMENT

AIDS has become the 'greatest obstacle to development', Dr Peter Piot, Executive Director, UNAIDS reported. Of the global population infected with HIV, 66 per cent live in sub-Saharan Africa. The facts are stark: the direct medical costs for each infected person are US$30 per day and average health spending in these countries is US$10 per day. The nature of the disease is such that health workers, the most needed of workers, are highly susceptible. Between 1999 and 2005 Botswana lost 17 per cent of its healthcare workers. In addition, the impact on farming can be appalling: Malawi will have an agricultural workforce 14 per cent smaller than it should have been by 2020.

Rostovian development will have to wait while so many countries in the world battle AIDS just to stand still in economic terms.

AFTERWORD

My wife can identify the department of origin of any French car by just two numbers on the registration plate. My American friend can name the emblem of each state of the USA and my Australian 'mate' can reel off the kings and queens of England. There was a time when to be 'good' at geography was to be good at remembering facts and figures, places and rivers. This so called 'capes and bays' geography started to disappear in the late 1960s when surveying both physical and human processes took over. More recently, in recognition of the sheer volume of such knowledge, a more descriptive subject focus has begun to emerge, the so-called 'qualitative revolution of geography'.

So what of the importance of geography in the twenty-first century? In the burgeoning world of the infosphere there is even more need for a holistic vision to make links between disparate discoveries in the world we live in. No one issue exists in isolation and indeed it is a source of exasperation among geographers that the media all too often does not lift its head from simple analysis of an isolated event to consider the much broader ramifications in the human and physical world combined. Of course, it is climate change where this is most presciently illustrated. As the failure of the Copenhagen summit of

2009 demonstrated, getting consensus between countries at different levels of economic development with different national priorities and with differing resources is possibly too big an ask. Witness subsequently how the climate-gate and glacier-gate furores threatened to undermine any future agreements. For every climate change debate there is another life-threatening issue: extreme flooding linked to El Niño linked to over-population in hazardous coastal and floodplain areas. So many of these issues require a holistic overview – and so lend themselves to geographers. Modern geography is about understanding the interlinkages both within and between the natural and human worlds. In addition, the subject is about using techniques such as geographical information systems (GIS) to help explain the complex in simple terms.

Why, therefore, is there a resistance to employing geographers in the media to add commentary and analysis? Perhaps it was the 'capes and bays' period of learning? Or just possibly some people do believe that geography is just about colouring in? Well, if all you got from this book is a realization that there's a lot more to the subject of geography than colouring in and blue questions in Trivial Pursuit, then we have achieved something.

FURTHER READING

Cool It: The Sceptical Environmentalist's Guide to Global Warming, by Bjorn Lomborg, Marshall Cavendish, 2009

An Appeal to Reason: A Cool Look at Global Warming, by Nigel Lawson, Duckworth, 2009

Six Degrees: Our Future on a Hotter Planet, by Mark Lynas, HarperPerennial, 2008

The Global Casino: An Introduction to Environmental Issues, by Nick Middleton, Hodder Education, 2008

Collapse: How Societies Choose to Fail or Survive, by Jared Diamond, Penguin, 2006

The Vanishing Face of Gaia: A Final Warning, by James Lovelock, Penguin, 2010

The Map That Changed the World: A Tale of Rocks, Ruin and Redemption, by Simon Winchester, Penguin, 2002

ACKNOWLEDGEMENTS

Thanks to my family for first nurturing it, Pete Bull and Chris Joseph for lighting the blue touch paper, Christelle and Fleuve for having to live with it.

Thanks also to David Woodroffe for his illustrations, and to Dominique Enright and Kate Inskip.

INDEX

Entries in *italics* denote illustrations